AUSTRALIS OSCAR 5

OWEN MACE

ADELAIDE
2017

AUSTRALIS OSCAR 5

The Story of how Melbourne University Students Built
Australia's First Satellite

ATF
PRESS

ADELAIDE
2017

National Library of Australia Cataloguing-in-Publication entry

Creator: Mace, Owen, author.
Title: Australis OSCAR 5 : the story of how Melbourne University
                students built Australia's first satellite / Owen Bullivant Mace.

ISBN: 9781925309805 (paperback)
                9781925309812 (hardback)
                9781925309829 (ebook : Kindle)
                9781925309836 (PDF)

Subjects: Melbourne University students.
 Artificial satellites--Australia.
 Artificial satellites--Launching.

An imprint of the ATF Ltd
PO Box 504 Hindmarsh
SA 5007
ABN 90 116 359 963
www.atfpress.com

Graphic Design & Layout: Lydia Paton
Font (Body text): Garamond (11.5pt)
Font (Heading text): Garamond (28pt), Futura LT (12pt)

To the young students who built and flew Australis, and to the memory of the late Geoff Thompson and Les Jenkins—they reached for the stars.

OWEN MACE

Acknowledgements

Australis was a team effort across the continents, first and foremost. Some are mentioned in this book and some, unfortunately, not. Thanks to each and very one of the Australis, WIA, OSCAR, AMSAT and NASA teams and to the many amateur radio operators throughout the world who listened to its beepings. Without each and every one, Australis OSCAR 5 would not have flown and not been the undoubted success that it was. Each person contributed something to the project and was part of its success. It seems unfair to name those I remember and leave out those whose names have slipped from my memory. They all contributed and thanks to them all.

Credit is due to Dr. Alice Gorman ('Dr. Space Junk') for her suggestion to write a book about Australis and to Anne Johnson for her documentary film that was my inspiration.

I must acknowledge my publisher, Hilary Regan and ATF Press, for taking on the task of steering me, who has never written a book before, through the publishing maze; for the patience of the copy-editor, Amy Lovat, for finding literally thousands of errors and changes necessary to make my initial manuscript readable and to Lydia Paton for her beautiful layout of the book. Thanks, too, to Pauline and Richard for their meticulous checking.

My English teachers deserve credit, too, for their patience with a student whose skill with the written word can only be judged by the 'rivers of red ink' that covered my essays at school.

And thanks to my beloved wife, Delia, for her decades of patience with her nerdy husband and for her wise counsel.

Owen Mace
ADELAIDE, AUGUST 2017

Plate 1a *This is a publicity photograph of the team taken in a Carlton alley very near the University of Melbourne. We are looking a good deal younger in the photo than today! In the top row, left to right, are: John, Paul, Steve, Richard (behind the antenna) and Geoff. Kneeling in the drain are Owen, Peter and Steve.*

Contents

What is this? Read on to find out...

Plate 1b *What is this? Read on to find out.*

Introduction

This is the story of the first satellite built in Australia, one that was built by students at the University of Melbourne in the 1960s and launched by NASA in 1970.

What were things like in the 1960s—the technology? Who were those students? And what about the satellite itself and its flight in space? Read on to find out.

We have been told that our achievement was remarkable and yet we did not think so at the time. We were simply following our interest in anything space; or to use today's terminology, our *passion*. Nevertheless, I think I can say that we all recall those times with pleasure and pride. We did achieve something unusual, although that was not our intention. We just wanted to do something that fascinated us, and the fact that no one in Australia had done anything like it before was of no concern or interest to us.

What is remarkable are the changes that have taken place since then. The developments in technology, especially electronics, are obvious and incredible. However, the ever more suffocating blanket of regulation now makes the attainment of such passions so much harder. But enough on such matters for now.

This all happened nearly fifty years ago, and some of our memories have dimmed a little, or rather have dimmed a lot, and there are even some instances where we disagree. Where there are differences, they are highlighted. For example, I have a clear memory that the amount of cash that we had available through donations and contributions from the Student Union and the Wireless Institute of Australia (WIA) was $400, whereas others are adamant that it was $1,400, a large sum for those days and especially for university students. There is general agreement among the team that the Student Union contributed $400. This was not unusual as the Union contributed to many, perhaps all, student clubs. The WIA contributed a like, or larger, amount. Newspapers record that the WIA's contribution was $400.

Does it matter now what the exact amount was? Probably not. Our activities and contributions to the project differed and perhaps I was never aware of the WIA contribution. Our memories are just different. This is a personal indulgence, recording my version of a time in the lives of a disparate group of mainly undergraduate students at the University of Melbourne from 1965 and as they progressed through graduation and, in some cases, postgraduate studies, to 1970 and beyond. We were bound together by a fascination with anything and everything relating to space. Not so much astronomy, but rather rockets, satellites and lunar probes.

The form of this book is based on the website—www.australis-oscar5.weebly.com—where I have recorded the story in far less detail.

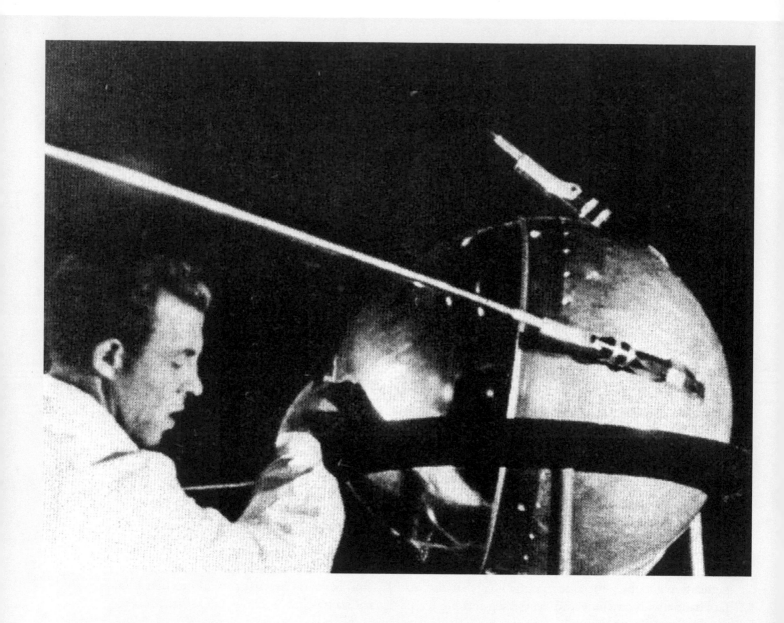

Plate 1.1 *Sputnik 1. Image Credit: NASA/Asif A. Siddiqi, from https://www.nasa.gov/multimedia/imagegallery/image_ feature_924.html* Accessed 31 July 2017.

CHAPTER 1
The Spark

'Pens down and onto the lawns,' said 'Frosty' Coleman.

Now Frosty was not someone to be trifled with, especially as he was the housemaster in a boys' boarding school, Geelong Grammar at Corio, and we his eleven-year-old charges. Not only was Frosty a somewhat severe man, to be not so much feared as respected, in a fearful sort of way. He had been a bomber pilot in World War II, so he was certainly respected by his charges for his wartime exploits, but also for the way he wielded his cane on boys who strayed from the designated path.

Those boys were born around the end of World War II and their parents lived through it. They were enthralled by wartime exploits, having read books telling the stories of the *Dam Busters*, Douglas Bader in *Reach for the Sky* and *The Great Escape*.

On this evening around 8pm, Frosty pointed to the lawn and playgrounds outside our study room for thirty or so boys. They trooped out, happy to be relieved of some study time and also a little fearful of what the reason might be. Perhaps it was something to do with Frosty's interest in the stars and astronomy? It was.

Once gathered on the playground grass on the cold, dark but clear night, Frosty commanded the boys in his bomber-pilot way to look up to the north west. Nothing but a thousand—no, a million—stars. And then something magical happened.

A spark of light was moving toward the group and in a couple of minutes it was overhead. Then it disappeared. That, Frosty explained, was an artificial earth satellite, the first ever. But not one of ours. It was the enemy's, the Soviet's. Would it attack? We wondered whether we were doomed. Possibly not, because it had a funny and unfamiliar name: Sputnik.[1]

I was one of those boys and that spark of light lit a lifelong fascination in me with anything and everything about space. There were others, too, who saw the same thing and were equally as excited by that little spark.

This is the story of where that spark and their imaginations led them—into space.
Richard's introduction to space was remarkably similar to mine. Here is his story.

* * *

---

1   Crikey, 'Sputnik 1: the night Australians watched the skies' at <https://www.crikey.com.au/2007/10/03/sputnik-1-the-night-australians-watched-the-skies/>. Accessed 31 July 2017.

Richard's Introduction to Space

My introduction to space research was remarkably similar, in fact almost identical, to Owen's. Like Owen, I was at boarding school when, on the night of 5 October 1957, our housemaster herded we Year 9 students up the hill to the Shrine War Memorial in Melbourne and told us to look up at the sky where we would see Sputnik, the first artificial satellite, which the Soviet Union had launched the day before. I suppose I had heard that they and the Americans had announced plans to orbit satellites during the International Geophysical Year, but the mere announcement had not piqued my interest.

But here was the real thing – a 'star' that moved quite quickly, it seemed, across the sky—an event that would have had its parallels in my grandparents being amazed at the invention of electricity, the Wright Brothers' first flight, television and the motor car.

I was thirteen at the time—half child, half man—with nowhere else to direct my boundless adolescent energy. While my boarding school friends were tinkering with model aeroplane engines, playing sport or thinking of girls (which was all we were allowed to do), I soaked up as much information as I possibly could about the Russian and American space programs. I had a small transistor radio that I would listen to wherever I could, for the latest space news—including in bed at night, pressed to my ear with the blankets pulled over my head.

I took this to the extreme. The United States Air Force began a program in 1959 called Discoverer (also referred to as Corona), which, as Owen says in Chapter Three, was, we thought, going to launch Australis in 1967. Discoverer was a spy satellite, with a capsule designed to be ejected from the parent craft and recovered as it parachuted to earth, by being snagged with a hook trailing from a slow moving piston engine aircraft. Through 1959 and 1960, they tried and tried to make it work, but 11 times it failed – the launch rocket did not get Discoverer into orbit, or the capsule was ejected the wrong way. The failures went on and on. Now, in those days, space was still big news and every launch was reported by the media. The success or failure of the capsule recovery was announced on the 12.30pm ABC News. On the day of each attempt, I would excuse myself from my Maths class (which was no great loss, as I was, unlike Owen and the others, quite hopeless at the subject), saying that I had to go to the toilet, where I produced my little radio and listened anxiously to the news. This happened frequently enough for the Maths teacher to take me aside and ask whether I was getting enough All Bran for breakfast.

At long last, on a wet August day in 1960, I was perched on the loo, awaiting the fate of Discoverer 13. The ABC News' grand piece of band music played (as it still does today), heralding the 12.30pm bulletin. The first item was all that I heard: 'The United States today recovered a capsule from an orbiting satellite. An Air Force aircraft failed to snag it as it parachuted toward the Pacific Ocean south of Hawaii, but a navy diver, dropped from a helicopter, located the capsule bobbing in the sea and it was winched aboard. President Eisenhower has hailed the feat as a major advance in space exploration', or words to that effect.

I was ecstatic—I whooped 'Yes!' very loudly. A boy in the next cubicle must have clearly heard me and my little secret was out.

Now, you might say that it was completely irrational for a then-sixteen-year-old to get excited about another country's space success, but that was the way I was. The space bug infected me that October night when I first saw Sputnik and it is with me still, sixty years later.

So far as I know, I was the only one at school, day boys or boarders, who was at all interested in space. I was regarded as a nerd, because once the initial excitement over Sputnik had faded away, for my school friends and the public generally, it was only to be rekindled in the early 1960s by the first human spaceflights, whereas I continued to be enthralled with each launch.

On my exeats, a couple of weekends a term when we boarders were allowed to go home, and on school holidays, I would build two stage rockets out of skyrockets that could then still be bought freely from fireworks shops in China Town, which occupied a couple of blocks in Little Bourke Street in Melbourne. The parallels with Owen's similar efforts are remarkable. I launched these space vehicles from the backyard at home, usually with less than satisfactory results; many fell into neighbouring gardens, spluttering and smoking. Luckily, our neighbours were sympathetic to 'the Tonkin kid' and his strange hobby. They called it a hobby to me it was an obsession.

* * *

We now return to the lead up to Sputnik 1 and events following it.

The International Geophysical Year (IGY, 1 July 1957 to 31 December 1958) was eighteen months of co-operative international scientific study of the Earth that had huge significance. Sixty-seven countries, including Cold War enemies The Soviet Union and America, contributed to the advancement of knowledge about the Earth that led to the discovery of mid-ocean ridges, and hence the theory of plate tectonics, as well as Arctic and Antarctic research stations. The Antarctic Treaty that committed countries to the peaceful and scientific uses of the Antarctic was a direct result of the IGY. Sputnik was launched during the IGY and its role was to explore the electron content of the ionosphere. There is a great deal of detailed information about Sputnik 1 in Wikipedia[2].

In the meantime, I scoured every newspaper for the slightest detail, and there was almost none. I recall one of the annual trips into 'town'—the central business district of Melbourne—with my grandmother, I saw headlines of the afternoon newspaper shrieking that there was now a second Sputnik and, evidently,

---

2   Wikipedia, 'Sputnik 1' at <https://en.wikipedia.org/wiki/Sputnik_1>. Accessed 31 July 2017.

a great deal bigger and heavier than the first one[3]. The Soviets had launched a second Sputnik[4] on 3 November 1957 and it carried a dog, Laika, which died soon after in space. How sad that the dog had to die for the Soviets to prove their superiority. Much later, we learned that she had died a rather cruel death soon after reaching orbit from overheating and carbon dioxide suffocation.

To much fanfare and publicity, America announced that their first satellite was to be launched soon after Sputnik 2. The launch was televised and, much to our dismay, the first stage rocket sank after a few seconds and disappeared in an enormous ball of flame. Dismay because it seemed that 'our team', the good guys, were being out-rocketed by the bad guys. Further adding to the dismal time was the fact that Sputnik 2 weighed around 500 kilograms compared to the Vanguard "grapefruit" that weighed a mere 1.4 kg. The Soviets seemed to be way ahead in the space race. What would become of us?

Though unintended, the success of Sputnik 2, as well as the very public televised failure of the American Vanguard launch later in the year, led to the 'space race' between the Soviet Union and America, much to the delight of those eager young observers.

By way of explanation, for boys like us, the world was indeed divided between the 'good guys'— Australians, Americans and the British, and the baddies—the Soviets and the peoples of the Union of Soviet Socialist Republic (USSR). Needless to say, the view was encouraged by the press and leaders of the country and my school. A few years later, in 1962, as I had progressed from Middle School to Senior School, the Cold War looked as if it might become seriously warm. The Soviet Premier, Nikita Kruschev, sent missiles to Cuba and they were able to deliver nuclear warheads to the US. Photos from US spy planes clearly showed the missiles on board ships headed for what President Kennedy called 'that Imprisoned Islands', and clearings in the Cuban jungle where they were to be based. President Kennedy called Kruschev's bluff and, after a few days – weeks it seemed – the missiles were withdrawn. It was kept secret at the time, but Kruschev insisted on what President Trump might call 'a deal' that US Jupiter missiles based in Turkey and aimed at the USSR would be removed, and they were a few months later.

Nevertheless, the risk of all-out nuclear war seemed to be high, even in my school. I clearly recall the tension and dread that I felt as I regularly looked skyward to see if there was a mushroom cloud. Mercifully, there never was.

There is a downside to all this rocketry and space stuff.

* * *

---

3   Wikipedia, 'Sputnik2' at <https://en.wikipedia.org/wiki/Sputnik_2>. Accessed 31 July 2017.
4   Crikey, 'Sputnik 2: the space age Australia never had' at <https://www.crikey.com.au/2007/10/03/sputnik-2-the-space-age-australia-never-had/>. Accessed 31 July 2017.

Arriving at Orientation Week at the University of Melbourne in February 1965 was a rather frightening moment for me as I realised that life beyond the sheltered walls of school had begun. The fear evaporated when I discovered the Melbourne University Astronautical Society (MUAS) stand. Here were a bunch of people who shared my interest in space and things that go there. I joined immediately.

* * *

Before beginning the story of MUAS however, we had better understand the times. To coin a huge understatement, they were rather different from today! In all sorts of ways, that is the key to the story.

FASHION DRAMA IN 3 ACTS

● With Light Fingers' Melbourne Cup finish one of the closest ever, racegoers had the added excitement of another Carnival drama. This was in the fashion field — the stars, English model Jean Shrimpton; Paris model Christine Borge. These pictures show the three acts — "The Shrimp" at right; Christine Borge, below.

THE DERBY: Casual shift-style — no stockings, no hat, no gloves, such a short skirt — that shocked Melbourne.

THE CUP: Conservative ensemble that saddened "The Shrimp," shown with Terence Stamp.

THE OAKS: Dress and buttoned jacket that was a happy compromise.

Plate 2.1 *In 1965 English model Jean Shrimpton shocked Melbourne race goers at the Racing Carnival when she wore miniskirts, no stockings, no hat (except at the Melbourne Cup) and no gloves. Such was convention at the time that this was news and a shock to Melbourne society. (Image Credit: Bauer Media Pty Limited / The Australian Women's Weekly November 1965)*

CHAPTER 2
No Boundaries—The Times, They are a' Changing

In this chapter, we will briefly look at the times in which the overseas space industry started and grew. Since then, many, many things have changed—obviously, technology has developed hugely, but also people's attitudes and government has changed enormously. Nevertheless, the 60s, and for that matter the 70s, were nothing if not exciting times for the young, at least, to live in.

We should not discount the effects of World War II, even well after its conclusion. Although the wartime government tried to keep news of attacks on Darwin and Broome from lowering people's morale, Sydney was attacked by midget submarines that passed the submarine nets protecting the Harbour and attacked shipping. As an aside, my father surveyed the torpedo test range in the Harbour, where torpedoes were test fired to measure their bias before being sent into action. The rationing that so much affected Britain was felt in Australia as well—I recall in the early 50s the Children's Hospital in Melbourne asking for donations of eggs for the children under its care.

The Cold War followed World War II, the Malayan Emergency and the Korean War. The Berlin Wall divided East and West Germany and the Cuban Missile Crisis nearly plunged the world into all-out nuclear war.

In many ways, the defining event (or series of events) of the period was the Vietnam War, or The American War as it is popularly known in Vietnam, which grew out of France's embarrassing defeat at Dien Bien Phu in Indochina, in what became Communist North Vietnam. Following the Treaty of Paris, the French exited Cambodia, Laos and Vietnam, which was partitioned into North and South Vietnam. France had asked the US to use nuclear weapons against the Viet Cong but, mercifully on advice from the UK, it did not. It is most unlikely (in my view) that it would have achieved anything other than universal condemnation and strengthened the determination of the Communist leaders.

The Vietnam War started with US advisers being sent to assist South Vietnam's army in 1955, with ever increasing numbers and eventually large numbers of American and Australian combat troops. For the record, and contrary to popular belief, Australia was asked to help South Vietnam—my father told me that he saw President Diem's letter to the Australian government requesting assistance.

The community's resistance to the war increased throughout the 1960s, culminating in civil disobedience and huge anti-war marches throughout the US and Australia. Not only was there opposition to the war itself, but also to the conscription of young Australian men based on their date of birth. My number was drawn but I was rejected because of asthma. Vietnam's war grew in intensity as more and more Americans and Australians were thrown into it and more and more were injured and died.

Among the worst decisions that the US military made was to allow and encourage the press to report freely on the war. With hazy black-and-white television reaching Australia's living rooms at the same time, graphic pictures of death, injury, misery and destruction were relayed directly into people's consciousness and consciences. With conscription and perhaps your neighbour's son being injured in Vietnam, the war was no longer something in a strange faraway land, but being played out in the living room and next door. It was not popular, and by 1973 the politicians had decided that it was not worth it and the war ended in ignominy for the US and Australia. I record with shame the fact of Australian citizens condemning soldiers, including conscripts, returning from Vietnam after risking their lives and doing what the government asked of them.

At the same time, the popular culture was developing—Johnny Ray was shocking adults in the 50s and 60s with his music and lyrics while delighting the youth, including me. I recall a concert he gave at the Sidney Myer Music Bowl in the 1950s that I couldn't attend although I did hear some of his songs from outside the arena. As popular music developed and expressed the thoughts and feelings of an optimistic generation born during and after the war, but never having really known the horrors of war, we began to question the norms and politics of the day.

It is reasonable to ask whether the war fuelled the development of popular culture or the growing popular culture fuelled the global Peace Movement and its opposition to the Vietnam War. I suspect that each grew from, and fuelled, the other. Whatever the causes and effects, opposition to the war, popular culture and questioning grew during the 1950s and 1960s. Things were changing and the 60s was an exciting time in Australia. Not just questioning and changing, but also growing in optimism and prosperity. Ordinary people were buying homes with electricity, running water and sewerage, cars, children's bicycles and televisions.

By the mid 1960s, the Beatles with their funny accents and lyrics that challenged our elders were engaging the youth as never before and encouraging us to further question. Even world leaders were no longer immune from being protested. I recall President Johnson's visit to Melbourne in 1966.

Huge numbers of students planned to protest Johnson's escalation of the Vietnam War and they gathered along the route of his motorcade shouting, 'LBJ, LBJ, how many kids have you killed today?' A banner on a balcony opposite the University apparently welcomed Johnson but was to be dropped to reveal a protest message as his car passed. The authorities bypassed Melbourne University at the last moment, leaving the banner missing its target. Nevertheless, two brothers later managed to throw a paint bomb at him, which missed but hit the unfortunately named Rufus Youngblood, a US secret service agent who thereafter was not very secret.

The peace movement, it seemed, was synonymous with drugs and free love. 'Recreational' drugs such as marijuana and LSD became readily available, especially for the young, as did sexual freedom, which came to a grinding halt in the early 1980s with the identification of AIDS and the HIV virus.

Another area where rapid changes were taking place was in clothing. No longer did we have to wear suits or sports jackets when at lectures or outside. 'Neat casual' was okay now, even jeans. But women's fashion was hardest hit. Conservative Melbourne was horrified when English model Jean Shrimpton wore a mini-skirt, with no hat, no gloves and no stockings in the Members' Enclosure at the 1965 Melbourne Cup (an international horse race held in Melbourne in November of each year). Heaven forbid, no stockings! The world was about to end . . . but it didn't. In fact, women were beginning to find their own freedom, not be tied down to home and children and even have their own careers. Germaine Greer's book, *The Female Eunuch*, appeared in 1970.

In areas that interested me more, cars were getting bigger, more powerful and somehow more strident with their acres of chrome, fins, V8 motors and bench seats. Some English and European cars seemed to move in a more peaceful direction with more rounded lines, such as the Saab and MG. Australian-made Holdens, of course, mirrored the American genre of a big flat bonnet and fins.

Much more important, however, was the march of technology, which built on the discoveries and engineering of the various wars when the military dominated research and technology development. Nuclear weapons and nuclear power spring to mind, but much more important strides were being taken in materials research and development. The obvious examples were transistor radios (but still there were many valve televisions) but behind the scenes the stage was being set for revolutions in materials of all kinds: plastics, steel, aluminium, concrete and even timber. The fundamental technology of the Global Positioning System (GPS)—super accurate clocks—was being investigated.

Computers were growing in power, size and numbers and there was even an IBM mainframe or two in Melbourne. These were huge monsters with flea-like computing power compared even with today's laptops (or even watches), let alone today's super computers. The University of Melbourne's IBM 7044 machine was installed in the early 1960s, the second in Melbourne after IBM Office's own machine. The central processing unit was five or six metres long and two high. Electrical connections between circuit boards were made by wrapping special wires around posts to make those electrical connections. The mat of wiring was as much as 10cm thick (international and Australian standards require a space between the number and its unit. The Australian standard). over much of the five-metre length of the backplane and two-metre height of the computer. Despite the tens – perhaps hundreds—of thousands of wire-wrapped connections, there was only one failure throughout its life. There were ten or so magnetic tape units and a miserable amount of memory (32,000 words, each 36 bits). I wrote my first program for it with punched cards, one card to a statement, in 1966. Unsurprisingly, it was a program to predict the direction of a satellite from a point on the Earth.

Despite what we think of today as antiquated, the IBM 7040/7044 and 7090 computers were built with transistors, which greatly reduced the size, electricity consumption and cooling requirements and so were a considerable step forward in computing over valve computers. There was much prestige in the University having a computer and so it was given its own cavernous room with glass windows for all to see and a considerable staff to tend to its every need. I would later see the next generation of IBM machines being developed, but that's for later in this story.

Jet aircraft in their current form and shape had made their entry but by no means was it commonplace for people to travel beyond capital cities. International travel was ridiculously expensive.

In many ways, the times were defined in terms of what was not available when compared to today. Television was, of course, analogue, black-and-white and fuzzy. The picture frequently became noisy without an adequate antenna and, in the early days, there were few recorded shows—it was all live broadcast. Certainly nothing from overseas, or even interstate, unless the recording tape was hand-carried to the broadcasting studio. Those who remember Graham Kennedy's *In Melbourne Tonight* show will continue to claim that it was the best evening entertainment ever produced.

Telephones were connected to their exchange by copper wires. Local calls were fixed price, unlike the UK system where all calls, even local ones, were timed and charged accordingly. Interstate calls in Australia were timed and expensive, while international calls were unthinkable. Indeed, the first international call that I had known was . . . well, that will have to wait until we reach that moment in this story. There were no personal computers, no internet, no world wide web, no emails and no social media. Life was simpler then.

On the other hand, President Kennedy had, just four years after Sputnik 1, committed the US to land a man on the moon by the end of the 1960s and return him safely. This was simply unthinkable at the time and many felt impossible. Nevertheless, with the goal in front of them, the US people accomplished a truly remarkable feat in a truly remarkable time frame and landed Neil Armstrong and Edwin Aldrin in July 1969. There were enough space-faring feats of daring almost monthly during the 1960s, as designs and ideas were tested and proved to keep this space nut thoroughly engaged. But not only did manned space flight advance enormously, so did unmanned civil and military satellites as weights, complexity and capabilities increased.

In some respects, though, the biggest difference then was the relative lack of regulation. Today, we can hardly move but for a huge mound of regulations, training and certification. A University of Melbourne 'cubesat' built, like Australis by students, was faced with a three-quarters-of-a-million dollar fee to get permission from the Australian government to be launched in the US! $750,000! If you wanted to do something in the 60s, you simply did it. Somehow, common sense ruled, not a vast and unapproachable (and often unobtainable) rule book. There were no boundaries, or at least few boundaries, it seemed. We were free.

Now, let's meet some of the men who came to be involved in Australis.

* * *

OWEN MACE

Like most Australians, I am the product of immigrants. On my father's side, they were from Warwickshire, Gloucestershire and Nottinghamshire in England. My father, Norman, must have been a particularly gifted and driven child. He related stories to me of evenings when the family could not afford coal for heating and he studied under a blanket for warmth and a candle for light. Winning scholarships to secondary school and to Jesus College, Cambridge University, he graduated in about 1930 with First Class Honours in Arts with a Mathematics major—certainly an odd combination it seems today. He was a first-class soccer player and played for the Cambridge team as well as for English county teams and for England in the Netherlands. I have the pocket from his Cambridge jacket signifying that he won a Cambridge Blue for football.

Once he had graduated, he needed a job. But what does a low-class man with an Honours degree in Mathematics do? He joined the Colonial Service and was sent far away so as not to disturb the natural order of the classes in England, as was his younger brother who followed the same course. Far away in Norman's case was Sarawak,[1] an independent state then under colonial rule on the western side of the island of Borneo. This was the land of head hunters and of the 'White Rajahs', the Brooke family. Norman became a surveyor and found himself carrying out survey work among the head hunters. After long service leave in England in 1939, the Japanese entered the war threatening, among other countries, Sarawak. Norman was eligible for leave but was unable to return to England, so he chose Australia where he joined General MacArthur's staff. In 1944, he found himself in Melbourne where he met his future wife, Nina May Bullivant, and they were married at Christ Church, South Yarra, on 30 September 1944. I was born eighteen months later.

My mother had a rather different background; also a family made good. On her mother's side, there were two first fleeters who arrived in 1788 as guests of His Majesty and provided accommodation at His Majesty's Hotel—they were convicts. One is said to have escaped the hangman's noose twice, was sentenced to transportation for life, was Jewish, became Australia's first policeman, was freed, became an innkeeper who defied the rum corps, returned to England and unsuccessfully sued Governor Bourke. Another, aged nineteen, was transported in 1832 for seven years for forging a promissory note for £200—a very large sum in the 1830s. Perhaps his rather pathetic note to his mother and sister, along with a bottle of laudanum (used to ease toothaches and commit suicide in those days) caused the judge to impose a very light sentence. He was very lucky!

Another branch, and this is the last convict, was a member of the expedition that attempted to found a settlement at Sorrento on Port Phillip before moving to Van Diemen's Land (Tasmania) and founding Hobart. The next generation moved to the Western District in Victoria where they ran huge sheep stations and were part of the development of Australia's wool industry, which was Australia's largest single export for a century and a half.

---

1   Wikipedia, 'Sarawak' at <https://en.wikipedia.org/wiki/Sarawak>. Accessed 31 July 2017.

Australian forces liberated Sarawak and Dad was posted to the capital, Kuching, to help rebuild the country. His wife and young son followed. I remember quite a few snapshots of my early life in Sarawak; among them, my father setting fire to the jungle, a cross section of the jungle through which the road passed and in which an orangutan and monkeys were visible, and my armah's (nurse's) home, a traditional long house that was decorated with human heads.

When the family returned to Australia in early 1951, Dad retired, but it wasn't long before he was recruited to the Joint Intelligence Bureau of the armed services—in those days the three services were run as separate government departments after Prime Minister Robert Menzies brought civilian control to the services following World War II. Certain functions, such as external intelligence, were run jointly. Internal intelligence was the province of the Australian Security Intelligence Organisation (ASIO), founded in 1949.

On return in 1951, I began primary school and then boarding school in Geelong where, in my first year, Sputnik sparked my interest in space. I clipped every newspaper article I could find and was given every book on space that could be found, including my favourite, The Conquest of Space by Chesley Bonestell, which is a collection of imaginative paintings of space scenes, as he imagined them in the 1950s.

* * *

Plate 2.2

In my later years at school, rockets were intriguing to me. I suppose it was little more than a boy's fascination with fire coupled with my existing interest in space. Rockets, after all, are a necessary precursor for a satellite. My early experiments were with multi-stage fireworks rockets – the ban on mere mortals using fireworks was a long way from being enacted. The idea was that, as one stage burned out, it lit the fuse of the next stage sitting above it. This seemed to work well, except that navigation was less than ideal. The rocket had usually turned over to point toward the ground as the second stage fired, resulting in a high-speed collision between the second stage and the ground, a house, or someone's precious flowers.

Later, a little research in the school library revealed that zinc dust and sulphur burn sufficiently rapidly that they make an adequate, low-tech and low-cost rocket fuel. Perhaps we could actually make a rocket. With a friend, Michael, we managed to obtain the zinc dust and sulphur fuel and oxidiser. An aluminium pipe about 25mm in diameter and perhaps 600mm long sufficed as the rocket body. A

Plate 2.2 *My father Norman, mother Nina and me soon after we'd arrived in Sarawak.*

nose cone was fashioned from, I think, aluminium, with a compartment for a parachute fashioned at the top. I knew that a nozzle was necessary, but why and how to design a suitable one was well beyond my knowledge. Never mind, a hole drilled in the aluminium plug at the bottom of the tube would have to be good enough.

We thought, 'this thing might be dangerous, so we ought to launch it somewhere away from people and buildings that might get in its way'. My father—and remember he was a spy—worked with Army officers and so he readily arranged an interview with Major Moustache, who had some influence at the Army training ground at Puckapunyal outside Melbourne. Listening to the improbable request from two pimply faced boys, the Major equally improbably agreed to their request to launch from a tank range at Puckapunyal.

And so it was that Dad drove us, the rocket and fuel to Puckapunyal where the rocket was set up and ignited by an electrical fuse, my first electrical design. With a loud report and lots and lots of smoke, the rocket instantly disappeared from our view—it was rising very rapidly upwards. Success! Despite an attempt to measure the height it rose to, we never saw it until it was on its return journey. Nevertheless, in the absence of data, we estimated that it got to 1,000 feet. Very scientific!

Plate 2.3

On examining the launch site, we noted that there was quite a lot of unburnt sulphur, indicating too much oxidiser and not enough fuel. The exhaust cloud was strangely yellow, too. Also, it must have taken off at a huge rate because the Masonite launch ramp[2] had been broken as the rocket hit the ramp. The nozzle remained at Puckapunyal, indicating that the attachment of the nozzle to the base of the tube was inadequate and so probably lessened the thrust. We learned from that flight that there is more to rockets than we had thought.

---

2   The German V1 rockets used launch ramps so, I reckoned, we should have one, too. Ours pointed not at Britain, but the sky.

Plate 2.3 *Owen, Nicholas, Nina and Delia in 1979.*

Sadly, there were no more flights, as final year exams beckoned.

* * *

While at school, my interests lay with what we today call the STEM subjects (Science, Technology, Engineering and Mathematics)—a rather corny acronym. Actually my subjects were Mathematics, Physics and Chemistry, but there's no decent acronym for that. I hated English as I just couldn't spell and so my essays were returned covered in rivers of red ink by my talented and long-suffering teachers. Nevertheless, something of their lessons must have sunk in because, once spelling and grammar checkers became available on the early computers in 1980, or thereabouts, I could write without the embarrassment of so much red ink, I began to enjoy writing and some of those learnings returned. Yes, I have since personally thanked one of my English teachers for his supreme patience.

Plate 2.4 *Owen, about four years old, in front of our home in Sarawak.*
Plate 2.5 *My armah, Emily, and me in Sarawak in about 1947.*

NSW UNI ATTENDS MORAT-ORIUM

· OVER 2000 OF US SET OFF FROM PADDO

SHOOTIN' REVOLUTIONARY THINGS

Plate 2.6

Another of my teachers at school taught Physics in a way that particularly appealed to me. For one thing, the Mathematics and Physics fitted together so well, which of course is unsurprising. So when I was mature enough to decide that I was too immature to go to University after my first matriculation year, Richard M. the Physics teacher in my second, repeated year of matriculation, persuaded me to switch from Science to study Electronic Engineering at University. What seminal advice for me, for the career that I was led into was quite beyond imagination.

* * *

Now let's see what happened to me, rather unsure of myself and my position in the world, when I entered tertiary education at the University of Melbourne.

Plate 2.6 *Vietnam Moratorium protesters, Tharunka (Kensington, NSW 1953 -2010) on 14 May 1970. (Image reproduced with the permission of Arc @ UNSW Ltd)*

Plate 3.1 *MUAS Club room; the 'rooftop garret'.*

CHAPTER 3
MUAS—Our Launching Pad

How did this disparate group of people meet?

Orientation Week, February 1965, saw the many University clubs jostling with one another to attract the attention of the wide-eyed freshmen, or first-year students, as we were then known. There were political clubs with choices between Young Liberal, Labour and, I think, even Communist clubs; there were sports clubs, perhaps even the lacrosse club that my grandfather played for in 1899 – I have a team photo with him, fresh faced as I undoubtedly was, and the top-hatted Professor Laurie, the club President. There were dramatic, amateur radio, debating and who knows what else societies. But the one that grabbed my attention was the Melbourne University Astronautical Society (MUAS), advertising that they were interested in satellites and space. That's for me, so I joined.

The Society had a meeting room, or club room, in the rooftops behind a lecture theatre of the 'Natural Philosophy' building. These days we'd call it Physics, but never mind; Natural Philosophy it was, housing lecture theatres, student laboratories for practical work, staff offices and laboratories. Somewhere hidden within the building, and I never saw it, was a cyclotron for accelerating electrons to high speeds and energies for research. When operating, the cyclotron played merry hell if we were trying to receive weak satellite signals by drowning them with electronic noise. The cyclotron was sold off decades later and used to produce x-rays for medical diagnosis and treatment.

Richard reached the University two years before me. Here is his story.

Plate 3.2

Plate 3.2 *Steve dutifully pointing MUAS' helix antenna on the roof of the Natural Philosophy building. Our 'rooftop garret' is among the roofs out of the picture to the left.*

At last, school was over for me – I was admitted to the University of Melbourne to study Law. Why was an eighteen-year-old space nut doing Law? As I said, I was hopeless at Maths, Physics and Chemistry. A careers person tested me over several hours and concluded that I should be a Librarian or a Lawyer. Law sounded more interesting. In those days, you didn't have to get a Year 12 score in the 90s to get into the Law Faculty—I only achieved a good Pass at Matriculation, as it was then called.

So there I was—Melbourne University, March 1962. Again, as with Owen, I wandered through the Orientation Week displays, and there was MUAS— the Melbourne University Astronautical Society. And there were guys (I don't recall any girls), surrounded by pictures of NASA and Soviet spacecraft, busily trying to recruit new members. They didn't have to try at all with me; I fell into their arms. There really were other space nuts like me—I had come home.

Plate 3.3

Plate 3.3 *Rod and Steve on the roof of the MUAS rooftop garret.*

Owen has described his introduction to the MUAS, three years later, with the same excitement. When I joined, the Society's main activity was showing space-related films that they had borrowed from the NASA office in Melbourne. We then moved on to the next logical step—tracking satellites, and we built a single, and then a quad helix antenna on the roof of the University's Physics building. The launch by NASA of the TIROS 8 weather satellite in December 1963 allowed us to receive pictures of clouds and the Earth below on a facsimile machine loaned by the Weather Bureau. Then, three years later, NASA's Applications Technology Satellite 1 in stationary orbit, 37,000 kilometres above the Pacific, provided spectacular coverage of a third of the planet.

But to the young space enthusiasts of the MUAS, the ultimate goal, surely, was the building of a satellite. That was how Australis started—one or more of us began talking about making our own satellite. No one had made one in Australia. That didn't matter to us; when people we talked to 'grown ups', they told us it couldn't be done, we ignored them. It has been said that human beings in their teens and early 20s appear to be programmed to pursue adventure, to take risks, to try things that their parents would consider dangerous—we had every reason to fail, but we didn't fear failure. In fact, I don't believe we even thought about it. People would ask, 'But is this project "official"?' To which we replied, 'No—we're just doing it'.

And so Australis was built, and this nerdy law student with the Buddy Holly glasses, although intellectually and practically incapable of designing or constructing any part of the satellite, nevertheless found administration and management tasks to do and contributed what he could.

* * *

Here's Peter's story:

My introduction to space was altogether different. Refusing to have anything to do with wearing glasses, I couldn't see Sputnik 1 when it was launched.

Around the time of the first Sputnik, I was interested in electronics and built electronic instruments, such as a valve oscilloscope, and transistor radios using the earliest semiconductors. I ventured into amateur radio and built my own receivers and transmitters and even managed to get my limited amateur license for the Very High Frequency (VHF) bands as I didn't want to learn Morse Code for the lower frequencies.

These hobbies kept me busy and, of course, I was interested in hearing about the space exploits of the USSR and USA. At University, I did a course in Astronomy. Around this time, computers were starting to appear and this started my interest in computing. After graduating, the first mini-computers started to appear and I built my own version of the Data General Nova computer. It was only through contact with the MUAS that I became interested in satellites.

* * *

The MUAS club room was reached through a lecture theatre, onto a flat roof, then up a steep ladder to enter the cosy 2x3x4m space with a single window[1]. In it, there was a desk along one wall and several chairs. On the table was a vast piece of equipment coloured battleship grey, for it was indeed a piece of cast-off Royal Australian Navy equipment. This, it was proudly announced, was the B40, and this was our radio receiver for listening to satellites. A capable receiver it was, too. Lots of knobs labelled with strange abbreviations, such as BFO, that I was to learn and understand in time.

The B40[2] was built for the Royal Navy and the Royal Australian Navy before it was cast off and found its way to a disposal shop where it was bought for the club. (Disposal shops that sold cast-off military equipment were common in Australian cities in the 1950s and 60s. They had all manner of wondrous equipment that excited amateur radio operators and MUAS members.) The B40s were manufactured in the 1950s by the Murphy Radio Ltd company in Welwyn Garden City in England, and each cost the lucky customers £500. This monster was 330mm wide, 483mm high, 406mm deep and weighed 46kg – no lightweight, but then it didn't need to be on a navy ship or submarine. The Radio Museum website describes it as being 'NOT Cathedral nor decorative'. You might not agree with the Cathedral part (it was big enough) but you certainly had to agree that it was not decorative. It was, however, functional.

For the technically minded, it was a High Frequency (HF) superheterodyne receiver, 640 kHz to 30.6 MHz in three bands with an unusual intermediate frequency of 500 kHz. Amazingly, it sported 14 thermionic valves, not the thousands of transistors that would be expected in modern receivers.

Unfortunately, however, there were certain limitations to this monster. For a start, we were reliant on a single receiver—there were no spare receivers, but the parts most likely to fail, the valves, were available.

Plate 3.4

1   How to gain the attention of someone in the club room after hours when the building was closed? Answer: a pair of wires from a bell in the club room running covertly down the outside wall to two nails. Bridging the nails with a coin rang a bell to alert the occupier to go down and open the main door.
2   Radio Museum, 'Naval Communications Receiver B40' at <http://www.radiomuseum.org/r/murphy_b40_b_40.html>. Accessed 31 July 2017.

Plate 3.4 *The Quad Helix attached to a lift well with Paul working on it.*

It was limited in other ways, too. The frequency coverage was not suitable for receiving signals from satellites and limited to amplitude modulation, but it did cover the lower frequency amateur radio bands, which turned out to be useful. Club members made 'pre-amps' that transformed the favoured satellite frequencies around 136 MHz down into the HF bands reached by the B40.

I also noticed that entering the room were some black cables that brought signals from the antenna, with pre-amp attached, back to the B40. But where was the antenna? Some thirty or more metres away on a large flat area of roof was our antenna, a clumsy contraption of aluminium pipes and wires standing on a post. It was steered manually by someone without much of a clue as to where the satellite was, and thus where to point it, and unable to listen to the whistles and beeps (or, more often, the electronic noise) from our B40 receiver. 'Move it a little to the left!' shouted across the rooftops to the man hanging onto the antenna handle. Not very satisfactory, so an intercom was installed to give the antenna man half a chance to find the source of our hoped-for satellite. Besides, he could then hear what the B40 was producing and let him decide how to maximise the signal.

It is unfair that the antenna should be called a 'contraption' or referred to as Heath Robinson. It was, in fact, a helix antenna with a not inconsiderable gain, and band width and beam width suitable for our uses. Although most satellite transmissions in those days were linearly polarised, unknown satellite attitude relative to the antenna and Faraday rotation in the ionosphere demanded an antenna that

Plate 3.5

Plates 3.5 *Youthful Paul and Richard working on the Quad Helix antenna.*

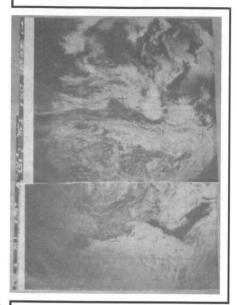

Plate 3.6

was agnostic to the angle of the polarised signal, so a circularly polarised antenna was the solution.

Even so, this was less than ideal. What was really needed was an antenna that could be controlled by the receiver operator in the club room. A design for a 'quad helix' antenna consisting of four of our existing antennas working together was needed. There followed a lot of discussion before the experts were satisfied and, needless to say, I contributed nothing to the conversation.

Now this was no minor construction. Each helix antenna was about three to four metres long and the design called for them to be mounted on a square and separated by about two-and-a-half metres. This meant a device about 4x4x3 metres that had to be pointed at the sky and steered to follow satellites while being entirely remotely controlled from the club room, which was about thirty metres away and not visible from it. An interesting engineering challenge for a group of mostly undergraduate students.

With higher gain, I learned, came narrower beam widths. (I was a first-year engineering student at this stage.) This meant that the new antenna had to be pointed more accurately at a satellite if it were to receive its signals and, in turn, a display of the pointing angles of the quad helix was required.

The design called for an H-frame to which each antenna was attached at the four points of the H. A substantial vertical pole held the H and the helices aloft. In order to point the helices at the satellite, there was a truck bearing at the centre of the H for the elevation axis and the vertical pole rotated in azimuth. This was a substantial piece of mechanical engineering, visible from around the University, and it had to withstand the worst that Melbourne weather could throw at it. Azimuth and elevation angles had to be read at the control station beside the B40.

Richard, the obvious leader of the group, organised weekend working bees to build the quad helix and was perhaps the hardest worker of the group. Nevertheless, we all contributed and, in really quite a short time, the antenna itself was built. Several nagging questions remained—where to mount it, how to drive the azimuth and elevation, and how to measure those 'Az' and 'El' angles.

The latter problem was fairly easily solved by the amateur radio operators, or 'hams' as they were often called, in the group. They knew that 'Synchros' were commonly used by those amateurs to measure

Plate 3.6 *APT pictures on wet fax paper showing the white antarctic coast in the bottom with the Ross Sea partly covered by cloud in the lower right corner. Moving northwards (and upwards), the second picture shows a huge low pressure system and part of the east coast of Australia on the right and highlighted by sun glint. Much of Australia is covered with cloud.*

antenna azimuth angle and were readily available from the disposal stores. A transmitter synchro on the antenna transmitted angle by way of three alternating current signals and a receiver synchro drove a shaft with pointer attached to display angle. Some experimentation was required to ensure that the three-phase AC signal wires were connected in the correct order so that the pointer followed the antenna in the right direction, but that was easily solved and the pointer calibrated. So that problem was out of the way.

Where to mount the quad helix? Well that was fairly easy, too—there was a lift shaft near to our existing 'single' helix antenna, so we could simply attach it to the brick shaft. The Physics department wouldn't mind, would they? Who knows, they were probably never asked, but they certainly never questioned it. Perhaps it was seen as a status symbol on top of their building. However, it did mean that the vertical, azimuth axis pole had to be well above the top of the lift shaft and so a steel pole was selected (yes, from disposals).[3]

Now for the more difficult problem: how to drive the antenna in azimuth and elevation? In order to explain our solution, dear reader, we must take a short diversion from the narrative.

Somehow, we found that aircraft propeller feathering motors were likely to be powerful enough to drive both azimuth and elevation These motors had an enormous reduction gearbox (9,000 to 1) with correspondingly enormous torque.

I soon found myself driving to Melbourne's airport, the Essendon Airport, and Ansett's maintenance department to collect two electric motors from a Convair aircraft. These were heavy beasts, meaning an even stronger vertical pole and a suitably strong mount for the azimuth motor. I now realise that there must have been considerable trust put in the azimuth drive as it bore the weight of the entire quad helix, something that it was definitely not designed to do. Thank you, Lockheed, for your excellent design. And thanks to Paul for manufacturing the heavy components, which he, Richard, myself and others assembled into the helix antennas ourselves.

The two feathering motors required a 24-volt supply at substantial current to drive them; barely a challenge for our hams. Underrated relays switched and reversed the voltage so that we could drive the axes in both directions and limit switches warned the operator in the club room of impending disaster. In the clubroom, a control panel was built with variable transformers (Variacs) to supply a variable voltage (via diodes, of course) to drive the azimuth and elevation motors.

Consider the further imposition on the building: the mass of cables between the antenna and clubroom: two cables, one each for the synchros; two heavy cables to carry the drive currents for the motors; four signal wires for the limit switches and intercom. We even had the foresight to include spare wires to replace those frazzled by short circuits.

---

3   Paul recalls worrying that, following windy nights, he might arrive the next day to find it draped over the building having caused major damage. He and I were relieved when it was finally dismantled.

After considerable effort in 1965 and 1966 and with the staff of the Physics Department ignoring our impositions on the building, our quad helix was ready to be powered for the first time. Variacs slowly turned up and, to our delight and amazement, the quad helix turned and nodded majestically, as appropriate for such a large beast. It worked! There was some argument and rework to 'phase' the synchros correctly; it can be somewhat disconcerting for the driver to be shown that the antenna is rotating clockwise when, in fact, it is anti-clockwise. Nevertheless, these minor problems were overcome quickly and it was operational. Now, to listen to some satellites!

* * *

The MUAS group was used to listening to various satellites using the single helix antenna. The Soviets had an active lunar program[4] even before President Kennedy announced his intention to see an American on the moon. Perhaps it was the Soviets' interest in the moon that led him to issue his bold challenge. Three Soviet spacecraft had been launched towards the moon—Luna-1 (January 1959) was the first spacecraft to escape the Earth's gravity, Luna-2 crashed onto the moon in September of the same year, and Luna-3 photographed the far side of the moon in, as I recall, singularly poor resolution.

In 1965, six lunar shots by the Soviets formed part of their program to match the US effort. All failed. The following year was rather more successful, with one achieving a soft landing and three orbiting. One of these was announced and tracked by MUAS, though I remain wondering if we heard the Luna or some terrestrial transmission. Whatever the true source of the signal, the press was notified of our intention to listen into the spacecraft's signals and it duly reported our 'success'. With all the space shots engaging the public's interest, we found a ready audience in the press, radio and television.

Even with 'our' satellites, that is American, we knew little if anything about how to decode the signals and so we could not gather any meaningful information from them, so we simply heard squeals and beeps and reported them to an interested press who published whatever we told them. If the beeps and burps happened to be a Soviet spacecraft, so be it.

How satisfying it would be if we could actually make something of the signals and decode, say, spacecraft temperature?

* * *

Richard recognised that the 1963 announcement of the launch of TIROS 8 (Television Infrared Observation Satellite) satellite[5] might be a satellite that MUAS could really see the data that it was transmitting. For the first time, a satellite was to transmit images of clouds below it in a form that possibly MUAS could receive and actually decode to produce recognisable pictures. Unfortunately, the press announcement did not include any information about how to turn the radio signal into a picture,

---

4   http://www.Sovietspaceweb.com/spacecraft_planetary_lunar.html. Accessed 31 July 2017.

5   NASA, 'The Television Infrared Observation Satellite Program (TIROS)' at <https://science.nasa.gov./missions/tiros>. Accessed 31 July 2017.

but the National Aeronautics and Space Administration (NASA) had a representative in Australia. Moreover, the Senior Scientific Representative was based in the city of Melbourne and a short tram ride from the University in Swanston Street. A telephone call yielded a meeting with the genial Willson Hunter, who provided us with the technical information we needed.

The frequency, 137 MHz, was within our capability, but the modulation was somewhat inconvenient—the 2.4 kHz subcarrier was frequency modulated (FM) onto the radio frequency (RF) carrier of 137 MHz and the faithful B40 could only demodulate amplitude modulated (AM) signals. A FM receiver costing hundreds of dollars was well beyond our means. What to do?

The answer was to dive into the innards of the B40 and pick up the intermediate frequency signal at 500 kHz and build an FM demodulator; a fairly straightforward task for the hams in the group. When I arrived at the club, there was a vigorous discussion about the success or otherwise of the demodulator. From my unknowing viewpoint, it seemed to work very well, even though a more modern receiver with FM capabilities would have been a bonus. The B40 continued to play its part for years[6].

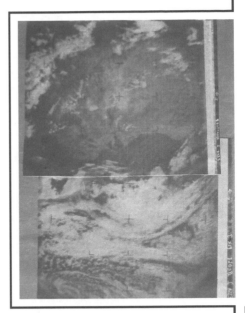

Plate 3.7

Plate 3.7 *The upper APT picture clearly shows the Great Australian Bight (lower centre) and the island of Tasmania (lower right side).*

The output of the FM demodulator was a 2.4 kHz subcarrier on which the scan lines of the image were amplitude modulated with line and frame synchronisation signals, rather like an analogue television signal, though much slower of course—two lines of the image per second. There were also bright and dark calibrations. This was a standard format, called Automatic Picture Transmission (APT), for meteorological organisations to send images to each other, typically on land lines, so equipment to display our cloud images was certainly available, but how to get hold of such a 'facsimile' machine to capture and record our images?

Perhaps the Bureau of Meteorology would be interested in helping us; after all, surely they would be interested in the images we produced. (It was commonly known as the Bureau then, not BOM, as it is now.) Through contacts with the Bureau, the loan of a high-quality, Swiss-made Kudelski Nagra tape recorder was arranged and so now we could record the 2.4 kHz audio signal, which was then taken to a 'fax' machine at the Bureau to be replayed. This at least proved that useful images could be received and provided to Bureau meteorologists, though I understand that there was some resistance to using images of clouds for their forecasts; after all, what could clouds really tell you about the weather? That resistance was soon overcome when they saw our pictures.

---

6   An Eddystone receiver, capable of demodulating FM signals, was later found for MUAS.

This was an unsatisfactory arrangement as there was no immediate confirmation that images were being received, so before long we received a fax machine on loan from the Bureau. This fax machine did not produce a crisp black-and-white image of a sheet of A4 paper. (In any case, Australia still used the British foolscap paper.) No, the image was burned electrostatically onto a damp, rather fragile, roll of paper (for want of a better word). That is to say, the paper was soaked in a conductive salt that turned dark when an electrical discharge was passed through it.

Imagine it: a machine that drove a pen across a roll of damp paper, discharging sparks to write the image as the roll slowly advanced. Once an image was complete, it was very carefully torn off the roll and then laid out to dry as the next image began to appear.

Despite the complex and obtuse path from RF signal to image, we could see clouds, coastlines and land features. We now had control of the reception of images from the antenna to the image itself.

Satellite APT transmissions continue, although other, much more sophisticated imaging and transmission techniques are used as well. Nowadays, a simple aerial, fairly cheap Very High Frequency (VHF) receiver and personal computer are used by enthusiasts throughout the world to capture images from later generations of APT-transmitting satellites. Furthermore, the images are in colour.

At last, we were doing something useful and real. Richard was employed by the Bureau to rise at some ridiculous time each morning in the dark, scooter to the University, let himself into the building, set the equipment going and receive the images from a 'pass' of Nimbus 1 and later satellites in the series. A Commonwealth car would appear near the building to collect the images and this always seemed to me to be rather cloak-and-dagger stuff, but probably reasonable in the circumstances and the times.

We, other MUAS members, were happy to contribute to this effort—we were, after

Plate 3.8

3.8

Plate 3.8 *Operator's console of the University of Melbourne's IBM 7044 mainframe computer.*

all, providing the Bureau with its only regular source of weather satellite pictures at the time. Until something went wrong.

I lived in a residential college nearby and early one morning there was a knock on my door. It was Richard explaining that the aerial was not working and there had been a fire. Now Richard, being a lawyer-in-training, was somewhat prone to theatrics—I'm sure he won't mind me saying so. 'Okay,' I said. 'Let's have a look.' Dressing and hurrying to the Natural Philosophy building, I discovered that indeed there had been a fire. That is, a relay or two and some wires had succumbed to the excessive currents imposed upon them. There was nothing that could be done then, so repairs had to wait for daylight and a trip to the electronics shop for more relays and wires – a morning's pictures lost. I'm sure that the relays were simply replaced; after all, they'd worked satisfactorily for months, despite being underrated, and indeed they continued to switch motor direction for years. 'Richard, don't suddenly reverse the motors—wait for a second or two before reversing.'

Although images were packed off to the Bureau, the process was to record the 2.4 kHz subcarrier on a professional quality Nagra brand tape recorder, which would then be replayed into the fax machine. This meant that the operator could focus on pointing the antenna correctly and coaxing signals from the receiver. During the replay, knobs and dials could be fiddled with to optimise the image quality, such as it was, and any errors could be corrected by simply replaying the tape again. This meant that multiple copies of the weather pictures could be made.

As we studied the images, we began to learn about the different types of clouds and what they meant. We could see cold fronts, a cyclone or two and the 'speckled' cloud of cold air behind a front. We were able to anticipate the evening's forecasts from the Bureau. The coastline of Australia, the Great Australian Bight, the Gulf of Carpentaria, the east and west coasts, even Port Phillip Bay were just as the maps had predicted! Occasionally images stretched as far as the unfamiliar Antarctic coast. What a success and what a treat for us techies.

Plate 3.9

Plate 3.9 *Array of magnetic tape drives of the IBM 7044 computer.*

One morning Richard excitedly reported to us that the first image of a pass that morning was totally blank. This was something that had not happened before. What had gone wrong with our equipment? He replayed the tape—no, just a steady subcarrier with no information on it. 'Wait, play it again, weren't there slight variations? Could that be an image? Unlikely, but let's see if we can get an image onto the fax.' After a number of replays and much knob twiddling, a very low contrast image appeared on the sodden fax paper. It was clear enough, however, to see a very unfamiliar coast with fjords, and was it snow and glaciers? Surely not! Well, it was indeed the last image taken by NASA's new experimental weather satellite, Nimbus 1, in the northern hemisphere before it entered the dark side of the Earth and ceased transmitting images. It was part of Greenland!

There had been a fault on board; the flash lamp had failed to erase the image from the detector and the image had slowly faded until the transmission sequence began again over Australia with the read-out of the image to l'il ole MUAS! Willson Hunter was contacted to find out what was going on. After sending tapes and images through him to the Nimbus project office, it was announced that we had seen the first convulsions of the dying satellite. Congratulations, Richard, for persisting with the difficult task of extracting the image from the very weak signal.

Plate 3.10

Plate 3.10 *Building the Quad Helix antenna.*

Something that came out of this project was to be of great significance and import to us, though we didn't recognise it at the time. The NASA representative, Willson Hunter, became a friend of MUAS and to me and others personally.

As a young engineer in 1945, Willson was in charge of the Icing research in the Wind Tunnels and Flight Division of the National Advisory Committee for Aeronatics (NACA)[7] at the Lewis Laboratory in Cleveland Ohio. He also worked for BF Goodrich and Co researching icing problems on aircraft.

Willson rose through the ranks and was a senior manager in the 1960s, prior to being appointed as NASA's representative in Australia. He was responsible for all NASA's tracking stations and operations in Australia and played critical parts in the successful reception of signals from the manned Apollo flights at Honeysuckle Creek, Parkes Radio Telescope and Tidbinbilla. There is a brief spot of Willson in a 1969 newsreel on Apollo 11, early in the movie 'The Dish'.

Willson's full contribution to AO5 in encouraging us, opening doors and greasing the wheels in NASA will probably never be fully known. Genial and urbane, he and his wife, Marj, were generous and encouraging friends. It was sad indeed to hear of his later demise.

* * *

This is all very well, pointing the quad helix and receiving weather pictures, but where to point the antenna? The satellites' orbits were precisely known and they appeared in two windows each day; several northbound orbits in the morning and several southbound twelve hours later. This is known as a sun synchronous orbit, having the advantage that illumination on the ground remains constant at any latitude, except for seasonal variations. This makes short-term comparisons easier as well as provides images at the same time each day for the weather forecasters. Nevertheless, a satellite 'pass' over a receiving station varies from day to day within each window, and so some form of prediction service is required. When and where to point the antenna—time, azimuth and elevation.

An interesting problem, how to predict the azimuth and elevation of a satellite. We knew that precise orbit parameters were published regularly by the US for their satellites, and some Soviet satellites. Perhaps that might help. Yes, the time and longitude of an ascending node was included. (The ascending node is the point in a satellite's orbit when it crosses the equator travelling northwards.) After a discussion with my father—and remember he had been a surveyor – he pointed me at an old surveying book of his. It's with me as I write this. He bought *The Text Book*

---

7   NACA was the precursor to and later became the National Aeronautics and Space Administration, NASA.

of Topographical and Geographical Surveying, published by His Majesty's Stationery Office in London, from Dymock's Book Arcade in Sydney on 25 January 1943 while on leave (and escape) from Sarawak.

The Text Book is pure Victorian England, having been edited by Colonel Sir Charles Close KBE, CB, CMG,[8] and a Fellow of the Royal Geographical Society, as Dad was then. Close's co-editor was also a colonel, H St JL Winterbottom, CMG, DSO. I imagine them to be pith helmeted and heavily moustachioed, but no doubt that's quite unfair. Wikipedia's account of Close's life shows him as having done survey work in Burma, Nigeria, the Cameroons and throughout Africa, including time as a commander of a surveying unit during the Boer War.

The text is a treasure house of minute details of survey instructions, methods, descriptions and formulas. There is all manner of useful information for surveyors, such as camp equipment, descriptions of surveying and analysis methods, tables and maps of the colonies where Close had worked. 'He was knighted in recognition of the Ordnance Survey's work during WW I when over 30 million maps were produced' according to Wikipedia. Close was Director General of the Ordnance Survey throughout the war.

Importantly, *The Text Book* had a section on 'Useful Trigonometrical Formulæ' (note the diphthong, '*æ*'), including equations for plane *and* spherical trigonometry. Immediately, once I saw those equations, the light bulb went on: knowing an ascending node, the altitude and period of the nearly circular orbit, I could calculate the location of the sub-satellite point at any moment and, hence, the azimuth and elevation of it from any point on the Earth. This was the way to provide antenna pointing information for the TIROS passes to an accuracy sufficient for the beam width of the quad helix.

In principle, anyway. The calculations with pencil paper and slide rule would be impossibly tedious to produce for several passes each day. No, there had to be a better way and there was—the University's brand spanking new IBM 7044 computer! It seems unbelievable these days and yet the IBM 7044 had 64,000 words of memory, or 192 kilobytes of magnetic core memory—remember that the very first IBM personal computers sported over three times this memory, 640 kbytes, and now many gigabytes are the norm.

Minor difficulty: I had neither the knowledge to program it, nor the permission to use it. No problem—remember these were days with 'no boundaries'—and permission was soon granted. A course in Fortran programming, some mathematics, coding sheets and time typing on the punched card machine soon produced a printout from the computer showing how many coding errors I had made. Correct the errors, re-punch the cards, submit the small deck of cards again on the next day, for there was only one 'run' per day, for the next batch of errors. Eventually, the errors were

---

8   Wikipedia, 'Charles Close' at <https://en.wikipedia.org/wiki/Charles_Close>. Accessed 31 July 2017.

eliminated and the program ran, producing antenna pointing information, but were the numbers correct? The next morning's passes proved that they were. We now had the means to predict when passes would occur and where to point the antenna. Furthermore, the accuracy of the process was accurate enough for a week or two's passes. Problem solved.

My recollection is that this was in 1966 when I was in my second year of Engineering studies at the University of Melbourne, but this does not align with the memories of others, nor with the dates for Tiros 8. Perhaps it was 1965. In any event, it was the start of a lifetime of engagement with computer hardware and software and even at the age of seventy in retirement, I program for my own satisfaction and recreation. But not Fortran, which then was a beast of a language!

In any event, with this introduction to computing satellite predictions, I was well placed for the next phase of the MUAS story.

* * *

For reasons I can no longer fathom, except perhaps for self-aggrandisement, we publicised our successes in the press, radio and television. This also turned out to be important for the next steps.

* * *

One day at our 'formal' weekly meeting discussing satellites, antennas, space, electronics and such like, somebody made the observation that we were successfully tracking satellites and obtaining weather pictures, so why not build one ourselves? No one has yet owned up to being that person. Then the discussion went along the lines of, well, why not, it would be fun, despite there being just twenty or so members of MUAS, some of whom like me were not able to contribute much technically.

The innocence of youth, especially in those times. In an article in the British *Classic and Sports Car* magazine about Adrian Squire,[9] founder of the short-lived Squire Car Manufacturing Company in Henley-on-Thames, Oxfordshire, the author observed:

> There is no energy source in the universe quite so powerful and intoxicating as youth. As people age, they tend to compromise, to settle, to reach no longer for the stars but for their spectacles—and then only to see restrictions and obstacles wherever they look. They rationalise. Youthful vigour (or arrogance, depending on your age) recognises no such bounds nor commonsense limitations. Nothing is impossible.

---

[9] Squire was unable to manufacture more than a few of his beautiful sports car and the company folded in 1936. He was killed in an air raid in 1940 at the Bristol Aircraft Company factory at Filton, Bristol, where he was working.

* * *

We will now meet some of those involved with Australis and read the stories of how they came to be involved. Later, we will see how they were involved.

* * *

Rod Tucker

Here's how Rod found the Melbourne University Astronautical Society:

'One cold evening in October 1957, I watched Sputnik cut across the Melbourne sky. I was hooked. Later, in the early 1960s, as a ham radio operator, I was inspired by the involvement of many hams around the world in tracking and sending signals to the early OSCAR amateur radio satellites launched in the US. In 1966, my first year as a student at the University of Melbourne, I visited the Astronautical Society clubrooms and watched as weather satellite pictures appeared on an early equivalent of a fax machine. There was a lot of activity in the Astronautical Society at the time, with the construction of AO-5 already underway. I quickly joined the Society and offered to assist with AO-5.'

* * *

Peter Hammer

Peter recalls his childhood years and later at the University of Melbourne, and how he became involved with Australis:

'My parents were born in Berlin in Germany: my father in 1900 and mother about ten years later. They endured the horrors of World War 1 and as a result my father could never eat pumpkin as that was the only vegetable available. He was a businessman and my mother did a PhD in Chemistry. They both had a Jewish background but never set foot in a synagogue. Nevertheless, my father found himself in a concentration camp in the late 1930s. Luckily his brother-in-law was Wilfred Burchett, who secured his release and both went to the United Kingdom, destitute.

Wilfred then went on to get a number of other people out of Germany. He was a humanitarian and later in life ran foul of the Australian Government with his stance against the Vietnam War and lost his passport as a result. Wilfred had a brother whose child is a very well-known Australian chef.

My parents emigrated to Australia under the 10-pound plan; 'ten pound poms', as they were rather disparagingly called. After arriving in Australia and a short stay in a camp, they established a business manufacturing adhesives. While my father ran the company, my mother produced all the manufacturing recipes while bringing up two boys.

My mother had been interested in photography for a long time in Germany. As well as taking photographs, she was particularly interested in the technical side of developing and enlarging. Her interest rubbed off on me and by the time the Olympic Games came to Melbourne in 1956, I was busy taking photos. My father was interested in psychology and he was an expert at graphology, assisting the police in a number of cases.'

Peter says that he had a happy childhood and did very well at his private school but his abilities fell away in sport. He left school at the end of Year 11 and enrolled in a Diploma of Applied Physics at the Royal Melbourne Institute of Technology; he was too young to enter University as there was then an age limit.

'I first became interested in electronics when I was doing a Diploma of Applied Physics at the Royal Melbourne Institute of Technology (now RMIT University). After graduating, I went straight into the second year of a Physics degree at Melbourne University. Here, I started getting interested in amateur radio and it wasn't long before I had my own call sign VK3ZPI. This was a limited licence restricted to the VHF and higher frequencies. I started building my own transmitters and receivers and often communicated on mobile amateur radio while driving around Melbourne.

'I gradually lost interest in Physics and switched to a Science degree, with an Electronics major. After graduating, I enrolled in a PhD program at the Physics RAAF Department, whose Professor was Vic Hopper. The undergraduate courses were run for air force cadets at the RAAF Academy, Point Cook, whereas the postgraduate school was attached to the University of Melbourne.

'My chosen field was Upper Atmosphere Physics, which is the study of the Earth's ionosphere starting at about 75km altitude. My PhD was mainly concerned with developing a new way of studying this region and it meant designing and building specialised transmitters and receivers. Altogether, this took several years of work. The resulting data was recorded on magnetic tape. In those days, the University had just acquired its IBM 7044 computer. As postgraduate students, we were allowed to book blocks of time of a few hours a day, running the machine after the daily 9-5 daytime shift. This saw me often running the computer late at night and having to cope with issues when things went wrong, like the line printer spewing out fanfold paper at rates of several metres/second or the card reader not wanting to work. Luckily there were backup systems.

'The head of my department was very much against spending time on anything but research so there were occasional meetings suggesting I stop all this "silly satellite stuff". To be fair, he was extremely dedicated to research and set an excellent example of single-minded focus on research. Needless to say, I still persevered with my PhD while building bits of satellite and playing with amateur radio. I also ran the ire of the senior electronic technician in the department, nicknamed "squirrel" for his way of locking everything up and making it hard for people designing and building electronics. (It didn't help that his nickname rhymed with his name!)'

JAN KING

Jan was born in Pontiac, Michigan, which was a General Motors town. As he remembers, 'I was in fifth grade the day Sputnik was launched. I was terribly interested in radios and later became a radio amateur. The President (Dwight "Ike" Einsehower) came on the air very quickly after the event and said that we have to win this race and I'm calling on all young people to become much more involved in space and in math and science. We need an aerospace community in the United States and this is a call for all those who might be interested to consider that in your career choice.'

The timing was perfect for Jan. He had been involved, with a friend, Bill, with radio amateurs building a satellite from his eleventh grade. General Motors offered his school time on their IBM 1620 computer and so Jan and Bill wrote a computer program in Fortran to predict when and where to point their antenna.

He and his mate, Bill, built a large antenna, not unlike the quad helix, on the roof of his home and tracked OSCAR 3 with it.

At twenty-one years of age, Jan graduated from University and was hired by NASA and assigned to the Goddard Centre outside of Washington, DC. Here, he was part of a team testing NASA satellites of all types and sizes up to huge astronomical observatories. As Jan says, 'I was like a kid in a candy store seeing all these weird scientific satellites every day . . . I had the bug.'

One day, a group of people asked him to join them building an amateur radio satellite. These guys were radio amateurs and had considerable technical and managerial seniority within NASA. They knew of Project OSCAR and that it was having difficulty getting agreement from the US Air Force for further launches.

* * *

MY VW

While a car is not quite the same thing as the people, this one figures throughout this story and so deserves a mention. I was exceptionally lucky—I owned a car, a VW beetle,[10] to be precise. It was probably one of the new 1966 models with larger windows and (wait for it) a larger engine—1,200 cc. For those who have not had the misfortune to ride in one, I simply say that the VW Käfer (German for 'beetle') was Adolf Hitler's idea for a people's car. Adolf rode in Mercedes Benz cars.

---

10   Wikipedia, 'Volkswagen Beetle' at <https://en.wikipedia.org/wiki/Volkswagen_Beetle>. Accessed 31 July 2017.

Rear engine, two-door, it was fully air-conditioned by opening the windows, had a reserve fuel tank that was operated by an out-of-reach lever and its rear wheels could fold under the car in certain situations. Nevertheless, it served Australis well and was the vehicle that we used to collect the Ansett prop pitch motors from Essendon airport.

Plate 4.1 *Peter's Very High Frequency (VHF) transmitter on the left and his High Frquency (HF) Transmitter on the right. They are sitting on computer printer fanfolder paper and hollerith cards on which is punched one line of a computer program.*

CHAPTER 4
Australis Electronics

We did not know enough to know that such a hare-brained idea as actually building and launching a satellite could never succeed, so we planned how we might go about it. We quickly realised that without a launch into space, building a satellite was pointless. Australia had not demonstrated a satellite launching capability at that stage and the only countries that had launched satellites were the Soviet Union and America. Obviously, the 'baddies' were not a consideration, even if we could speak their language, so an American launch was the only feasible option. Furthermore, our ham members knew that the California-based Orbiting Satellite Carrying Amateur Radio (OSCAR) organisation had already flown four satellites.

The Project OSCAR Inc. was formed in 1960 by amateur radio operators in California who were members of the TRW Radio Club of Redondo Beach. (TRW was an electronics company specialising in defence electronics.) Many worked at TRW, other defence companies and the Foothills College. Many were also members of the teams that built and flew the Discoverer or Corona missions. These were highly classified spy satellites that photographed the ground beneath them on 70mm film, which was returned to Earth in a return capsule. Believe it or not, the return capsule was supposed to be caught by a cargo plane trailing a rope in the hope of snagging the parachute risers. Unsurprisingly, the capture process was extremely difficult and not initially very successful. The return capsule was also designed to remain afloat for two days to allow recovery from the sea by the US Navy or anyone else passing by. There were 144 Corona satellites launched, of which 102 returned usable photographs.

OSCAR 1 was launched on 12 December 1961. It was no coincidence that many of the OSCAR people worked on the Corona program. In any event, space was found for a small payload on the Agena second stage from which it was ejected, presumably so that it would not interfere with the Corona satellite. It carried a 140 milliwatt (140mW) transmitter in the two-metre amateur radio band at 144.983 megahertz (144.983 MHz) and transmitted the hams' greeting, HI, in Morse code[1] (four dots followed by two dots) that was received by well over 500 amateurs in twenty-eight countries over the three weeks that the batteries lasted. This was a simple, low-cost satellite—the ejection spring came from a local store and cost less than two US dollars. This was just over four years after the first Sputnik.

---

1   Morse Code was devised by Samuel Morse in about 1837 for transmitting characters using a code consisting of dots and dashes, thereby building up text. For radio transmission, the dots and dashes are represented by pulsing a transmitter, or audio tones, on for short or long periods. From Marconi's first radio transmissions in the 1890s, Morse Code was used for long distance radio communication. By the year 2000 it had all but disappeared in favour of other means of communicating text.

This flight was quite an amazing achievement, not just because it was the first satellite to be ejected from another or that it was the first amateur radio satellite. How on Earth (literally) did Project OSCAR persuade the spy agency (the Central Intelligence Agency) and the US Air Force to fly a civilian satellite? The naysayers no doubt would have argued that it might compromise the success of the mission and its security. On the other hand, it helped that some of the proponents worked on the Corona project. In all likelihood, senior managers were amateurs also and this would have helped as well.

There is also another aspect. Amateur radio was a far more popular pastime and was more visible than it is today. While today amateurs are frequently the first to make contact from a disaster zone (such as Cyclone Tracy[2] in Darwin on Christmas Day 1974 and more recent cyclones and extreme weather events), their work is rarely acknowledged. Of course, in the 60s, amateurs normally made their own receivers and valve transmitters as well as using them to communicate. As the decades passed, however, commercial receivers and later transmitters became available to amateurs and fewer amateurs built their own equipment. With the internet, computers, mobile phones and tablets, the ease of communication reduced the allure of talking to someone on the other side of the globe on a noisy single sideband channel[3].

Nevertheless, in the 60s, there were large numbers of amateur radio operators in the US and many in senior technical and management positions.

 Plate 4.2

2   Wikipedia, 'Cyclone Tracy' at <https://en.wikipedia.org/wiki/Cyclone_Tracy>. Accessed 31 July 2017.
3   Single sideband is a modulation technique that reduces the amount of bandwidth needed compared with amplitude modulation and so improves the received signal strength. The disadvantage is that unless the receiver is tuned to the transmit frequency exactly, the speech is distorted.

Plate 4.2 *The VHF transmitter during construction. Note the ferrite tuning coils.*

OSCAR 1 was a pioneer in many ways[4]. The use of HI as identification, two-metre signals and with springs became something of a standard.

OSCAR 2, launched on 2 June 1962 on another Corona mission as a secondary payload, or, less politely, as ballast. It was similar to OSCAR 1 except that it was a rectangular shape at 300×250×120 mm, weighing 10kg with a single monopole antenna. The transmitter power was lowered to extend battery life and a measurement of its internal temperature was transmitted. Its external surface was changed to lower the internal temperature compared to the previous OSCAR. It transmitted for 18 days until the battery was exhausted.

OSCAR 3, weighing in at 16.3kg, was launched on 9 March 1965 and carried a very different payload, a linear transponder that relayed signals to extend the range of two-metre communications over thousands of miles. Over 1,000 amateurs communicated using the transponder and received the beacon signals.

OSCAR 4 was planned for a geostationary orbit when launched on 12 December 1965 with a payload similar to its predecessor, but it reached only a geostationary transfer orbit of 33,000km apogee (the highest point in the orbit) and 161km (lowest point). Its linear transponder received signals on 144 MHz and re-transmitted them on 432 MHz. The transmit power was three watts and it carried the first direct satellite communication between the US and the USSR before it re-entered the atmosphere on 16 March 1966.

The Project OSCAR people were visionaries who set their sights high and achieved a number of firsts. Amazing achievements less than ten years after Sputnik – and amateurs at that!

* * *

Well, we thought, if they can do it, so can we. In fact, perhaps the OSCAR people could organise a launch for us? The address of OSCAR headquarters was found – no internet then, remember – and a letter dispatched asking them if we built a satellite, would they organise a launch for us. A reply arrived barely two weeks later. Given that a week was the delivery time for an international letter then, our request must have been replied to immediately. The answer was yes, they would launch it if we built it. The challenge had been laid!

We now had to think carefully about what we wanted to fly. Clearly, it had to be a satellite that would interest amateur radio operators and, besides, some of our members were amateurs themselves. We made contact with the Wireless Institute of Australia, the WIA, who, with their members, supported us magnificently throughout the project.

---

[4]   Wikipedia, 'OSCAR 1' at <https://en.wikipedia.org/wiki/OSCAR_1>. Accessed 31 July 2017. See similarly for the other OSCAR satellites. Also: Andreas Bilsing, *OSCAR-1 Launched 50 Years Ago* (online). At <http://www.arrl.org/files/file/Technology/Bilsing.pdf>. Accessed 31 July 2017.

A name was found—Australis—to signify the southern hemisphere and relate to Australia. Project Australis was born with its headquarters in the rooftop garret of the Natural Philosophy building in the University of Melbourne, postal address care of the Student's Union and telephone number… well, who remembers these days?

So, in general terms, there must be something for amateurs in the Australis satellite. But what? We knew, of course, of the transponders that Project OSCAR had flown. The ultimate for an amateur satellite, the 'gold standard', therefore, would be a transponder that received amateur transmissions on one frequency in the two-metre amateur band, amplified it and re-transmitted on another frequency in the same amateur band. In this way, a two-metre radio transmission, normally limited to not much more than line of sight, can be beamed to a satellite transponder that is hundreds, even a thousand, kilometres away. The signal is then re-transmitted to the receiving station a further thousand kilometres away. The two stations might even be on opposite sides of an ocean[5].

Technically, a two-metre transponder is a much harder device to design and build than a simple two-metre beacon. The reason for the technical difficulty is that the transmitter would be in close physical and frequency proximity to the receiver. Keeping the receiver deaf to the nearby transmitted signals while at the same time being highly sensitive to its own receiving frequency is a difficult task further complicated by the harsh space environment that was largely unknown to us[6].

The OSCAR people, whose profession was flying satellites, well understood the space environment and how to build electronics to operate in space. So it was decided to 'cut our teeth' on a simple satellite,

---

5   This is a standard technique for marine communications, which are on frequencies close to the two-metre amateur band. A transponder may be sited on a high headland or hill to extend the range of marine communications for many tens of kilometres, or more when line of sight communication for a small recreational vessel might be just ten kilometres.

6   Modern marine transponders accomplish this, but the ironmongery required to adequately filter the transmission signal from the receiver was far too large for Australis.

Plate 4.3

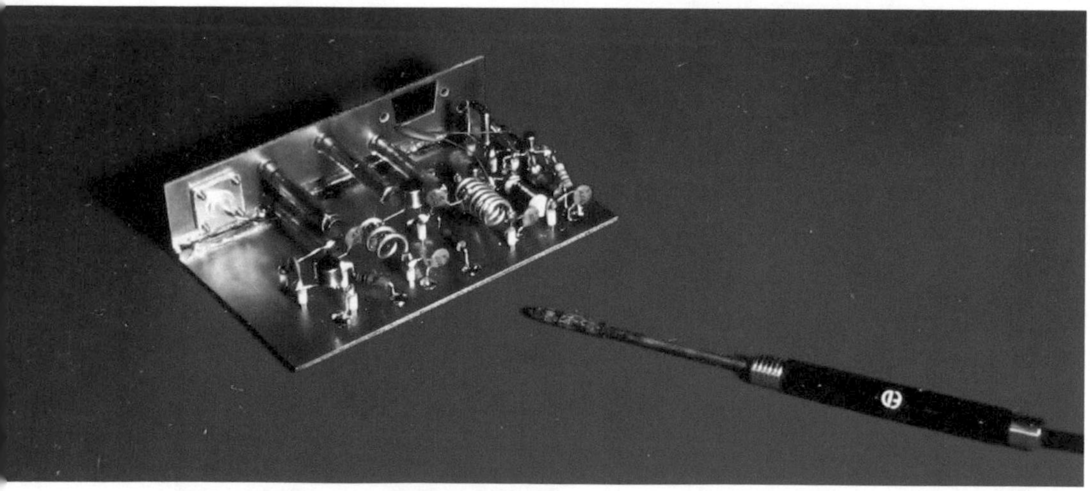

Plate 4.3 *A prototype HF transmitter showing the tuning capacitors mounted on the vertical panel.*

42   AUSTRALIS OSCAR 5

but with the view to future satellites carrying transponders, the ultimate amateur radio satellite. While seemingly a simple decision, it took some discussion to reach it.

To be useful, any future transponder satellite that we dreamed of had to be stabilised in orbit; that is, it must not tumble in orbit, otherwise signals would fade as the satellite's transmission pattern became unfavourable for the amateur's ground transmitting station or the receiving station. This had been a problem with the earlier OSCAR transponder satellites. Therefore, a stabilisation system was required and a means of measuring the satellite's orientation was also needed.

As Australis was to be a demonstration or test satellite, showing the way for future amateur radio satellites, we decided to fly two transmitters, a two-metre transmitter as on previous OSCARs, and a ten-metre transmitter in the amateur band, a first for any satellite at the time of our design. The transmitters must not just say 'HI', they must carry useful information, such as battery voltage and current drain. If the ten-metre (High Frequency, HF) transmitter was to be useful, it had to transmit at a considerably higher power than the 100 mW two-metre transmitter in order to overcome the greater path loss at ten metres. In fact, the ten-metre transmitter nominal power was 250 mW—high power indeed for a battery-powered satellite of the size we envisaged. This led to the idea of a command system that could control the 'high power' HF transmitter.

Add a battery, a means of maintaining the temperature of the electronics, and separation system, a structure, and we have the fifteen sub-systems of Australis, which were:

1. A stabilisation system to reduce any residual tumbling in orbit.

2. An orientation system to ensure that Australis' antennas are in a direction favourable for receiving the signals on the ground with no, or minimal, fading.

3. A two-metre, 50 mW transmitter, amplitude modulated by the telemetry signal.

4. A ten-metre, 250 mW transmitter, also amplitude modulated by the telemetry signal.

5. A command system consisting of a receiver and decoder to turn the ten-metre transmitter on and off.

6. Antennas for the two transmitters and the command receiver.

7. A 'HI keyer' to generate HI in Morse Code.

8. A telemetry system to generate the modulation signal for the two transmitters.

9. A battery to power the electronics.

10. A separation system to eject Australis from the launch rocket, including a master on/off switch.

11. A structure, including thermal control. Thermal control was a matter outside our control and finally designed by NASA. Nevertheless, provision had to be made for a temperature control layer to be added to Australis' outer surface.

Without amateurs listening to Australis and reporting the signals they received, the project was worthless, except perhaps as an ego trip for us students. Accordingly, a ground segment was recognised as consisting of:

12. A network of amateurs to receive and report the signals they heard.

13. A means of generating and distributing predictions on when and where to look for Australis.

14. Command stations to send command signals to control the HF transmitter.

15. An analysis centre to analyse reception reports from receiving stations.

Although we did not express the design decisions in this way, this was a system design that clearly delineated Australis' components, functions and interfaces, even if it was not formally documented.

One aspect that still puzzles me is how we arrived at the size and mass specification. Perhaps we were thinking that it should be about the same size as OSCAR 3. I was never aware of any specific need to minimise size, mass or even power consumption.

At the same time, we needed a reception station at the University to ensure that we had up-to-date information on the flight. This was readily achieved with the equipment we already had for our weather satellite reception, but we had no transmitter capable of commanding Australis, so this was designated to Bill Naughton in country Victoria. Bill had substantial VHF antennas (and possibly slightly over the limit transmitter power in the two-metre band), had participated in some interesting amateur radio experiments, such as bouncing a signal off the moon, and was keen to help.

Designers and builders of the components volunteered, or perhaps, more accurately, were volunteered, so that we had a team working toward the goal of building Australis. Peter took on the two transmitters and the HI keyer, David the telemetry system and command decoder, Paul (with help from Geoff) took

Plate 4.4

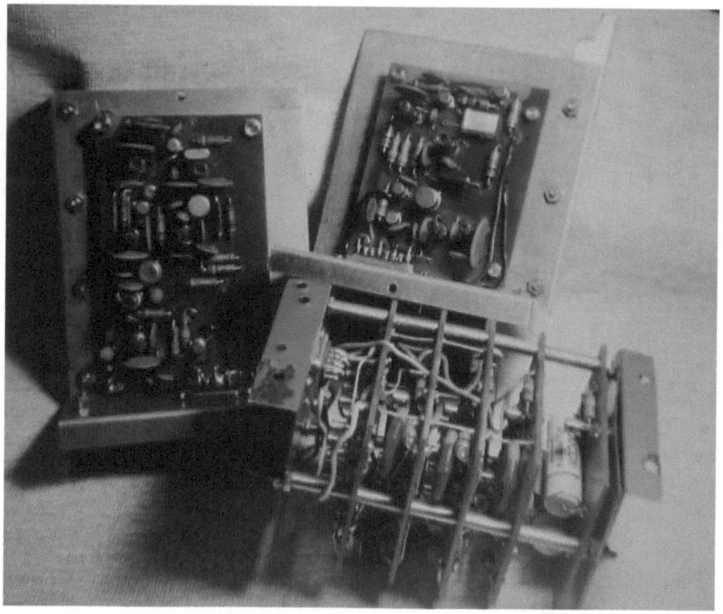

Plate 4.4 *The two transmitters and HI keyer.*

on the mechanical aspects, and Steve, with help from me, the orientation and stabilisation system. The only person who was not a student was Les Jenkins, who built the command receiver. He was a keen amateur who often allowed us into his radio shack to hold conferences with OSCAR using his HF radio. Crammed into his shack in the bottom of his garden and surrounded by glowing valves, dials, knobs and lamps, I found the distorted voices from the single sideband HF radio difficult to understand, but we managed to communicate with OSCAR people successfully.

Richard and I took on the task of providing supplies for the team. We understood that the space environment for the electronics was quite different from the earthbound environment. We knew that Australis' internal temperature might easily become very high or very low—we had no way of knowing at the time—so designers set themselves a wide operating temperature range. We knew that radiation was likely to be a problem; the Van Allen radiation belts that surround the Earth[7] had been discovered but we had no understanding of the problem. In the end, the best quality components we could get hold of were used with the exception of the space-qualified transistors, which we managed to obtain.

There was virtually no budget for the project, so almost everything had to be donated. Richard and I, therefore, undertook a publicity campaign in the press, radio and television. At every opportunity, and as we shall see there were quite a number, press releases were issued and our faces appeared in front of the public at a surprising frequency. The outcome was that on telephoning the manager of a company that provided something we needed, we were known and even recognised when we visited. Certainly, I cannot recall an occasion, even interstate, when we were sent away without presenting our case. Nor, for that matter, can I remember a refusal, such was the interest in space and the interest that we had created in our satellite project.

By the end of 1966, we had a team busily working on their contribution to the project and, considering that each of us had our studies to attend to, the team worked surprisingly well.

We had no concept of project management but the role was necessary to ensure that steady progress was made. Quite unreasonably, Richard and I pushed and cajoled those "volunteers" to design, build and test their contributions.

* * *

We will now briefly look at the details of each of these sub-systems and stories behind them. While the orientation and stabilisation sub-system was central to Australis' design, we begin with the electronics and in the next chapter describe it and the structure, batteries, ejection and master power switch systems.

* * *

---

7   *Wikipedia, 'Van Allen Radiation Belt' at <https://en.wikipedia.org/wiki/Van_Allen_radiation_belt>*. Accessed 31 July 2017.

TRANSMITTERS

As we read in the previous chapter, Peter was a member of the University of Melbourne Amateur Radio Club and experienced in building radio frequency transmitters and receivers. When I worked with him later, he seemed to have a happy knack of understanding what to me were complex RF circuits. The amateur radio club and MUAS met together from time to time to discuss their hobbies and it was during one of these meetings when he met Richard and me. Also at one of those meetings, the idea of building a satellite was raised and he was interested.

To begin with, Peter undertook to build the two transmitters and the HI keyer. Peter was one of the two PhD students in our group. He had started his research, which entailed studying the ionosphere with others in the Physics RAAF Department using radio methods. His professor was the kindly and dedicated Vic Hopper. Notably, under him was Jean Laby, daughter of an early professor of Natural Philosophy at the University and author of the Kaye and Laby logarithm and trigonometric tables that mathematics students used every day. Vic Hopper's department, that I was to join later, was physically and organisationally separate from the 'main' Physics department that occupied the Natural Philosophy building where our antennas and meeting room were located. Physics RAAF provided teaching staff for the RAAF Academy at RAAF Point Cook—more on that later.

Peter's early work on his studies required him to build transmitters and receivers to probe the various layers of the ionosphere and he had an affinity with radio frequency electronics. He built the electronics that worked beautifully. I worked with another man later in life who, likewise, had a great affinity with electronics. Whereas I had to calculate component values, he somehow simply knew the right value and, invariably, he was correct. Just as well, as he didn't know how to calculate them! Returning to Peter, he was the ideal person to build the two transmitters. He had built similar transmitters before and understood the technology very well.

The High Frequency (HF) transmitter operated in the ten-metre amateur radio band at 29.450 MHz and was amplitude modulated by the audio signal from the Telemetry System. The output power was nominally 250 mW, which was considered to be a heavy drain on the battery and so it could be commanded on and off by the Command System. There were a mere three silicon transistors in the transmitter, an oscillator at the transmit frequency, a modulator and a class C power amplifier. However, the key to the transmitter was not so much the transistors but the filtering to ensure that spurious signals were suppressed and the desired 29 MHz signal was delivered to the antenna at the required power level.

Like all the electronics, it had to operate in a wide, but unknown, temperature range. While we were able to specify the battery voltage, the battery would discharge eventually. Accordingly, all the electronics should operate to as low a battery voltage as reasonably possible, the transmitters all the while observing the specifications for spurious signals so as not to interfere with receivers on other frequencies. Interference with other electronics within Australis was also a concern. This was no easy task, especially for the transmitters, and you can see the many filter coils and capacitors in the photograph.

As with the other electronics, prototypes were manufactured and tested to satisfy ourselves as much as possible that they would work correctly in flight. But more about testing later. The prototypes were important for another reason. All the flight components had to be accommodated within the satellite structure so the prototypes were useful for understanding how they would be packed into Australis' mechanical structure—no three-dimensional computer modelling then!

Being the RF guru he was, Peter also built the two-metre, Very High Frequency (VHF) transmitter. Though conceptually simpler not having to be commanded, it was to remain turned on throughout the flight; the much higher frequency was a challenge, if only because silicon transistors had reached those frequencies only relatively recently. On the other hand, the power output was nominally 50 milliwatts at 144.050 MHz. Like the HF transmitter, it was amplitude modulated by the Telemetry System.

The design was more complex, however, with four silicon transistors. The 36 MHz crystal oscillator provided a signal at one-quarter of the transmit frequency, which was doubled and then doubled again before being amplified in the final stage. The first frequency doubler was also modulated by the telemetry system. Again, the photograph of the prototype clearly shows the filtering coils and capacitors.

It is worth discussing two concerns that we had. First, the value of the inductance of the coils used in the filters was directly related to their ability to remove unwanted frequencies as the resonance frequency is set by the inductance and capacitance values. Changes in either change the resonant frequency and thus the filtering performance. It was understood that there were two reasons why the inductance and capacitance values would change. One cause was temperature changes, which cause (among other things) dimensional changes in the components. For the inductance coils in particular, the inductance is determined by the physical dimensions of the coil and so as the metal expands and contracts with temperature changes, the inductance changes. Likewise, the filter capacitors also change values. This was a considerable design challenge for Peter and something that was discussed at our lunchtime meetings in the club's rooftop room.

Plate 4.5

Plate 4.5 *Hi Keyer boards before assembly.*

Peter built the VHF transmitter for AO5 using ferrite core inductors to allow the inductance to be tuned to the correct value, whereas the HF transmitter used open coil inductors and high quality adjustable capacitors to tune the circuits. The transmitters were then 'potted', that is, filled with epoxy resin, to ensure nothing moved and hence frequencies remained as constant as we could make them.

Both transmitters and the command receiver used quartz crystal resonators, known simply as 'crystals', to set the transmit frequency. However, quartz crystals, and hence the transmit and receiver frequencies, do not remain exactly as their name plate frequency as their temperature changes—they have a temperature coefficient of frequency. Different types of crystals have different temperature coefficients and some even have a point where the coefficient is zero but all change frequency slightly with large swings in temperature. It all depends on the way in which they are cut from the original piece of quartz. Care was taken to specify the appropriate crystal type, or cut, for the frequency and temperature range we anticipated for Australis. Here was another task for the "scroungers" of components.

As if he had not enough to do, Peter also built the HI keyer. Unlike his radio frequency work, digital logic was something that I could understand—like the stabilisation system, it was a subject that I had just learned about in lectures. The HI keyer was somewhat more complex than the transmitters, but

Plate 4.6 *Peter's Hi Keyer.*

Plate 4.7 *The Command Receiver.*

only in the sense that there were some fourteen transistors and about thirty-five resistors. The design consisted of a four Hz multivibrator (oscillator) followed by four dividers that reduced the frequency to one-quarter Hz plus four logic gates to produce the Morse Code modulation signal of dot, dot, dot, dot, pause, dot, dot for HI.

The earliest microelectronic components consisting of two or more transistors and other electronic components in a single package, or chip, were appearing around this time and they were Resistor-Transistor-Logic (RTL) logic chips. It was decided not to use them as we did not know whether any were space qualified. In fact, the logic form that Peter chose for the HI keyer was RTL but implemented only with discrete transistors and resistors. Throughout the years, more robust forms of logic have been developed and RTL has followed the dinosaurs into oblivion.

Unlike his transmitters where layout was critical, the layout of the HI keyer was not and a more important criterion was to minimise the volume (and power) that it occupied. Peter chose a "cordwood" construction, which was a common technique then, especially in space and military applications. Cordwood layers small circuit boards in three dimensions with wires connecting the various layers. The advantage is efficient use of volume but manufacture is difficult and repair near impossible, unless the fault happens to be on one of the outer layers. Nevertheless, the HI keyer was built up to a block of circuit boards, transistors, resistors, capacitors, wires, nuts and bolts.

It was decided for some reason that the HI keyer should also be 'potted'. I can only surmise that we thought it would be more robust that way. It fell to me to pot it, which I did. I can reveal here for the first time that I included the name of the University college where I resided from a letterhead with the epoxy resin potting. A small part of Trinity College made it into space!

Plate 4.8 *Dave's Command Decoder boards before assembly. Note the large inductors and capacitors for the ON, OFF and ENABLE resonant audio circuits in the centre upper row.*

* * *

TELEMETRY SUBSYSTEM

The next electronic sub-system was the telemetry encoder, which delivered temperature, etc. information to the ground by modulating the two transmitters. The first consideration was what information we wanted. One of the important questions we were seeking an answer to was whether a satellite could be stabilised and oriented in orbit and so we needed to know how Australis was moving once it reached orbit and how that movement, or tumbling, changed with time.

I recall a lunch meeting where we wondered how we could possibly measure the tumbling. Richard, recalling instruments on various satellites, including the Apollo spacecraft (and especially the moon lander), suggested a star tracker. Now a conventional star tracker is a telescope with a light detector and motors to turn it to focus on the chosen star—this was much more complex than the rest of Australis. Nevertheless, his suggestion led to the notion of semiconductor light sensors that we expected to record the transition that the sensor viewed from the blackness of space to the brightness of a sunlit Earth. We later discovered that this was exactly how the weather satellites we were tracking monitored their deliberate rotations.

Three 'horizon sensors' were deemed necessary to understand Australis' tumbling, although in truth we had no idea how we would be able to understand the tumbling motion from the horizon sensor signals, but that was a problem for later. An interesting development occurred in flight, however. Not only were the horizon sensors able to detect space to Earth transitions, which they did very easily, but variations in light were seen as the sensors' view passed over the Earth. These were clouds – our own very crude weather satellite, but this was to be some years later.

So, three horizon sensors were to be buried in tubes to restrict their view and shade them from extraneous light. This led to a concern for the potential for damage and destruction should a sensor see sunlight directly. A balance had to be struck between a reasonable field of view for the horizon sensors and a narrow field to reduce the risk of damage from the sun. This was a risk that we had to accept but, again, this was something we had to wait for some years before we learned whether the risk was worth taking. In fact, no damage occurred during the flight.

We now had four channels of telemetry decided; HI was the first telemetry channel and used as an identification and synchronisation channel, for without the clearly identifiable HI greeting, it may have been difficult to identify the other telemetry channels. Then there were to be the three horizon sensors.

We knew that maintaining the temperature of a satellite within a range suitable for the electronics is critical. We have discussed frequency stability of the transmitters and receivers but changes in other electronic components with temperature also affect their quantitative value and hence the way they perform, so temperature of the internal parts of Australis was a major consideration for the design. Accordingly, two temperature sensors (which deliberately have large and deterministic temperature coefficients), internal and skin temperature, were specified. Finally, battery voltage and current measurements were added, making a total of seven telemetry channels plus HI. (Any power of two is relatively easy to achieve in binary logic electronics.) The eight telemetry channels were:

1. HI identification.
2. Battery current drain.
3. X axis horizon sensor.
4. Battery voltage.
5. Y axis horizon sensor.
6. Internal temperature.
7. Z axis horizon sensor.
8. Skin temperature.

The Telemetry System connected each of these eight telemetry channels in sequence to an audio oscillator whose frequency varied with the voltage of the sensor connected to it. The output of the audio oscillator modulated the two transmitters. However, being amplitude modulated, the transmitters transmitted more power, and hence drew more current from the battery when the modulation voltage was high, and so the transmitters were modulated by signals with opposite phase so that, while one was drawing maximum power, the other was drawing minimum current, thereby smoothing the overall current drain on the satellite's battery.

Each channel was sampled for 6.5 seconds for a 52-second telemetry cycle. The audio frequencies ranged from 400 Hz to about 3 kHz. For each of the battery voltage, current and temperature telemetry channels, there was a calibration graph provided in the Users' Guide to convert the audio frequency into the parameter being measured. More on the Users' Guide later.

Minimising temperature variations was also an issue for the telemetry system since changes in temperature caused changes in electronic component values that in turn changed the audio frequency. Design choices minimised temperature coefficients of the critical electronic components.

* * *

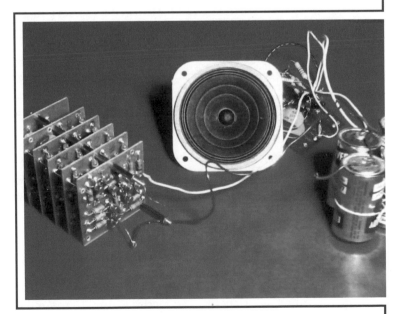

Plates 4.9 *HI keyer being tested.*

Command System

Recall that the ten-metre HF transmitter was designed to radiate five times more power than the two-metre VHF transmitter and hence it would draw more from the battery. It was felt that a system that would turn the HF transmitter on and off would not only allow us to manage the battery current drain, and hence battery lifetime, but also demonstrate for the first time that control of an amateur satellite could be achieved. The two commands required were straightforward—turn the HF transmitter on and turn it off—but its implementation was not simple.

Conceptually the commands would simply consist of two audio tones carried on a two-metre signal to turn the transmitter on and a different pair of tones to turn it off. To simplify, one of the tones, known as the Enable tone, was shared between the two commands, the others being the On and Off tones. The Command Decoder used the three audio tones to determine the state of a bistable flip flop that controlled the transmit state of the HF Transmitter.

There was concern that noise from the command receiver may activate the commands and so it was necessary to ensure that only a narrow range of audio frequencies transmitted for sufficient time would be required to command. The problem of maintaining frequency stability of the three audio filters in space over the expected temperature range was significant for the technology of the day, which was narrow band (technically, a high Q factor) resonant circuits with large, and temperature sensitive, inductors and capacitors. Dave managed to achieve these demanding requirements with another cordwood module.

The only designer and builder of Australis who was not a student was Les who built the complex command receiver. Technically, it was a double conversion, superheterodyne receiver with a pre-amplifier, two intermediate frequencies, amplitude modulation detector and automatic gain control. If complexity is measured by the number of transistors, it was relatively complex; there were twelve.

The frequencies were deliberately not published in the Users' Guide to ensure that valid commands were only sent by authorised stations but I can reveal them here. The receiver operated at 147.850 MHz in the two-metre amateur band. The ON command required the Enable tone of 2,960 Hz and the ON tone of 3,494 Hz to be transmitted simultaneously for several seconds. The OFF command required the Enable tone and the OFF tone of 3,226 Hz to be transmitted. The bandwidth of each of the command tone audio filters was 30 Hz.

Other than the master power switch, these were the electronic sub-systems on Australis.

Before turning to the orientation and stabilisation system, as well as the mechanical aspects of Australis, it is worthwhile discussing the methods used to make the electronics for Australis. The standard circuit board material used in industry in those days was bakelite, about 1.6mm thick and they were supplied coated with a thin film of copper. The boards were drilled and then, using photographic processes, the layout of the desired electrically conducting tracks was printed onto the board, which was then chemically etched to leave the track pattern. The track layout, which printed a material that was resistant to the etching material, was called 'resist'. After etching away the unwanted copper, the desired tracks connecting the various electrical components remained.

As poor students, we didn't have the resources to use those expensive production techniques. For the AO5 spacecraft, we used fibreglass for its superior strength and we painted the track pattern on the copper with nail varnish, which was our 'resist' material. The boards were then etched in ferric chloride to dissolve away the copper where there was no nail varnish. It was not uncommon for people using this technique to return home with holes in their clothes from the acidic ferric chloride. Goodness only knows what the fumes did to people's lungs. There were lots of things to go wrong, especially the etching, and often boards had to be remade and processes repeated.

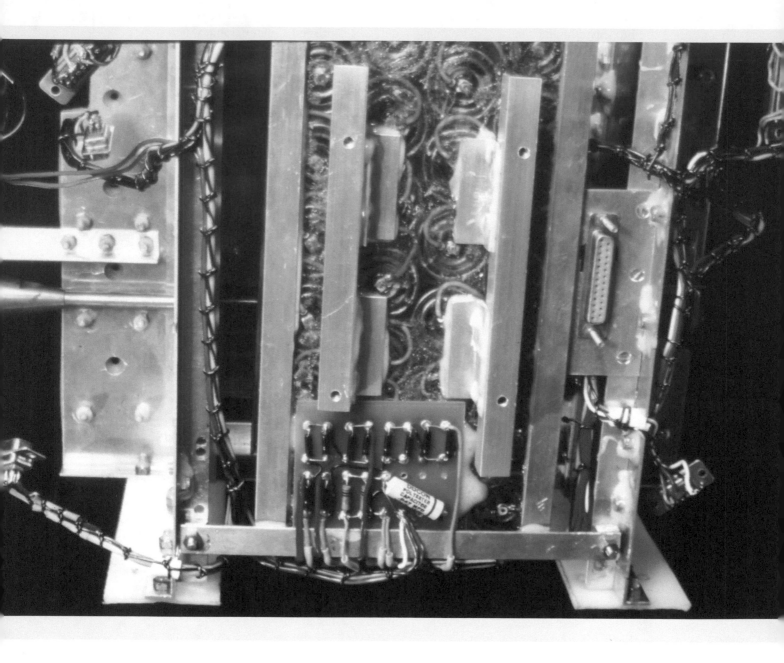

Plate 5.1 *Partially assembled satellite. The orientation magnet is clearly visible on the left, the batteries are in the centre of the satellite and the isolation diodes at the bottom of the picture. Notice also Paul's meticulous dressing of the wiring loom. The mountings for the upper panel of the satellite are glued to the batteries.*

CHAPTER 5

Australis Mechanicals, Bending the Metal

Perhaps a better title for this chapter would be Australis' Non-Electronic Systems – a bit of a mouthful, but more descriptive.

We begin with the Stabilisation and Orientation Systems. We expected that Australis may be placed into orbit and ejected from the launch rocket with a spin. Recall that it is desirable for a transponder satellite, and indeed any satellite transmitting radio signals, to have a reasonably steady signal at the receiver so as to avoid 'fades' where data is lost or compromised. This is as true today as it was in Australis' time, notwithstanding modern digital communications, perhaps even more so when a weak signal can cause complete loss of digital data, whereas a weak analogue signal may only cause loss of measurement accuracy.

Magnetic Attitude Stabilisation System (MASS)

Most large, modern satellites are stabilised and oriented in three directions using a complex set of sensors, reaction wheels and gas jets. This was out of the question for Australis, if for no other reason than we simply could not afford the electrical power necessary to run such a system. No, the stabilisation system had to consume minimal electrical power, or ideally none. Fortunately, the Earth's magnetic field, although weak, afforded us a means of crudely stabilising and orienting the satellite.

There are, therefore, two parts to the subsystem—a means of reducing any residual spin of a satellite as it enters orbit and another means of orienting it in a direction suitable for the antennas. The sub-system was called the Magnetic Attitude Stabilisation System, MASS, consisting of orientation and stabilisation components.

Stabilisation Component

Equally fortunate was that Steve and I, as electrical engineering students, had recently attended Physics lectures where the physics of magnetic hysteresis had been taught. When magnetic materials are subject to changing magnetic fields, energy is absorbed by the material as the atoms (or rather the magnetic fields of the atoms) within the material are re-oriented by changes in the magnetic field. In some materials, used for electrical transformers, for example, this energy is small, but in others much larger. Could a material with a sufficiently large 'hysteresis loop', as it is known technically, absorb the rotational energy of Australis? The size of the hysteresis loop is a measure of the energy absorbed by the magnetic material for each reversal of the magnetic field; so a large hysteresis loop, absorbing a relatively large amount of energy, is desirable for this task.

After consulting some technical books in the University's library, Steve reckoned that it might just be possible to markedly reduce any satellite spin to a reasonable value in a few weeks. Any spinning or rotating could never be totally eliminated because, as the speed of the spinning was reduced, so the number of magnetic field reversals, and the rate of absorbing rotational energy drops. Nevertheless, reducing the rate to, say, one rotation or fade per minute would make communication far more feasible.

Eddy currents in Australis' aluminium skin due to the changing Earth's magnetic field as the satellite rotated would also absorb energy and contribute to reducing the spin. Unfortunately, we were unable to calculate the likely contribution of this energy absorption mechanism.

In addition, we had no idea what the rate of rotation might be. Launch rockets typically rotate about their longitudinal axis to help stabilise the direction that it is pointed. Also, when Australis was ejected from the launch rocket, any imbalance in the ejection springs would add a further rotation, about a different axis. We had to live with the notion that Australis would be spinning when placed in orbit at a rate of perhaps a rotation every few seconds. It scarcely needs saying that this made the design of a stabilisation system and prediction of its performance difficult at best. Furthermore, at the time we had only a vague understanding of the likely mechanical properties of the satellite.

Nevertheless, we sought out a material with as high a magnetic hysteresis loop as we could find and measured the loop. We found an alloy consisting of about fifty per cent iron and fifty per cent nickel with a very high hysteresis loop and made by Allegheny-Ludlum Manufacturing in the US. A local company, Brown and Dureau, would import it for us at a cost of US$35, with us additionally paying for the freight from Dunkirk, New York, and customs clearance.

There was one further problem, however. A web page on Alloy 4750 states that '. . . in order to achieve the maximum softness and optimum electrical and magnetic properties, Alloy 4750 should be annealed in a dry hydrogen atmosphere at 2150°F for 2-4 hours, followed by a furnace cool at 100-200°F per hour down to 800°F and at any rate thereafter'.[1] Finding someone willing to heat hydrogen to high temperatures would be a challenge! Nevertheless, that brave someone was found at the Engineering Faculty of Monash University and the wires were duly soaked in a hot hydrogen atmosphere for the due time and allowed to cool at the due rate. The challenges of building a satellite! (I wonder if such facilities are even available today in Australia and if there is anyone willing to undertake the task nowadays.)

Following the annealing, there was a memorable afternoon as we set off to the student's electronics laboratory in the Engineering Faculty at Melbourne University to make the measurement of the hysteresis loop. Arriving at the laboratory with the material in the form of wires about 3mm diameter, an irate and insistent Senior Lecturer Jim P. demanded that I not enter wearing thongs—imagine the danger present when measuring a magnetic hysteresis loop when wearing thongs! (The voltages were low.) Nevertheless, the rules were the rules and I trudged off to change into something more to his liking, and in accordance with the rules. Eventually, the hysteresis loop was displayed on an

---

1   Aircraft Materials, 'Nickel Iron Alloy 4750, Technical Data Sheet' at <https://www.aircraftmaterials.com/data/electronic/4750.html>. Accessed 31 July 2017.

oscilloscope and the numbers recorded and analysed. This is not a difficult measurement to make (I've done it a few times since) but it was a novelty then, and a satisfying one, when done for the first time.

The upshot was that yes, with a handful of these wires, the likely spin of the imagined Australis could be reduced to a reasonable value in a reasonable time. With all the unknowns, any quantitative analysis and estimate of the spin-down time was guess work. It would work. Thanks, Jim, for letting us use the lab.

After all this challenging and very interesting experience, we had our stabilisation system—a bundle of innocuous wires. Better than any lecture or practical class at University, we learned and thoroughly understood the principles of magnetic materials and hysteresis loops.

Significantly and most importantly, the wires required no electrical power for them to do their job; they were passive.

ORIENTATION COMPONENT

The ideal orientation system would orient Australis in a fixed direction relative to the Earth, so that the various antennas could be arranged to maximise signals toward the Earth. As we've seen, an active orientation system was utterly impractical for us. We needed another passive system and, once again, the Earth's magnetic field offered the solution.

As any child knows, two magnets attract and align themselves. The Earth's field is one of those magnets, so all we had to do was to place a magnet in Australis and, hey presto, Australis would be aligned with the Earth's magnetic field. Since the Earth's field is very roughly parallel with the Earth's surface over much of the populated regions (but not at the poles), here was an easy solution.

Not quite so quickly. How big a magnet is required to follow the Earth's magnetic field, especially given that the field dips down at the Earth's poles and a satellite tends to do a 180-degree tumble as it passes over each pole?

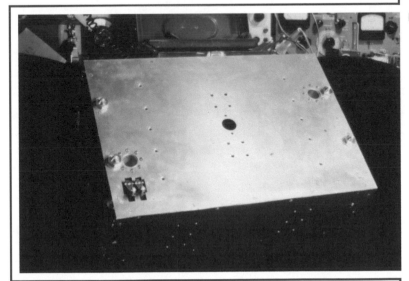

Plate 5.2

Plate 5.2 *The bottom plate of Australis showing Rod's four locating pins, two holes for the Henderson springs and the centre hole for the HF antenna mount.*

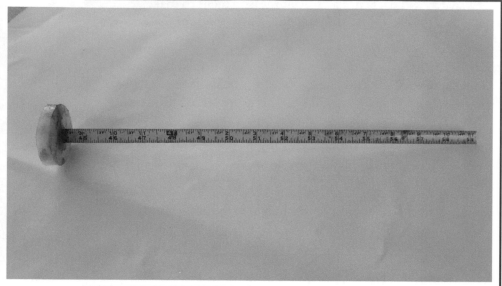

Plate 5.3

Furthermore, we had introduced a permanent magnetic field, which may affect the hysteresis of the stabilisation wires. These were serious considerations that had to be understood and dealt with. More delving into technical books dealing with magnetic fields and, in fact, much of the theory is contained in Physics and Mathematics that was well beyond our understanding at that stage of our University careers—vector calculus and the like. Nevertheless, Steve was able to convince himself that magnetic orientation was feasible.

The Rola company in Melbourne manufactured audio loud speakers. Perhaps they would be willing to supply us with magnets, and they did.

As previously mentioned, we had little understanding of how Australis would be spinning as it left the launch rocket.

Ejection Springs

On advice from OSCAR, we needed to eject Australis from the launch rocket. Further, had we remained attached to the rocket, the opportunities for mounting our three antennas would have been very much reduced, and in any event, the usefulness of a satellite periodically shadowed by the rocket would be very limited. The simplest approach was to use a pair of springs to push Australis away from the rocket. This required the rocket to have a mechanism to hold Australis in place before and during the launch, and then release it—but that would be the responsibility of the rocket people. We simply had to provide the springs and a locality mechanism.

The springs had to be matched; that is, each of the springs had to have the same force characteristics as they were compressed and released so that Australis would be pushed evenly into space with a minimum of spin imparted by the ejection. Persuading the Henderson Federal Spring factory to supply them, which it did, is a story related in the next chapter.

Plate 5.3 *A prototype of the antenna mount made by Rod. The antenna really was a tape rule!*

ANTENNAS

Three antennas were needed, one each for the two transmitters and one for the command receiver. The two two-metre VHF antennas were to be tuned monopoles with the structure of Australis acting as some sort of ground plane, whereas the single ten-metre HF antenna was a tuned dipole with the satellite structure itself in the middle of the antenna. The two VHF antennas were estimated to be about 475mm (19 inches) long and the HF antenna roughly ten times as long, with each arm being about 1,300mm. It was unlikely that any rocket would be able (or the organisation willing) to accommodate a satellite that, with antennas, measured at least 2.5x1m, so the antennas had to be folded into a more compact package.

A simple solution was suggested – flexible carpenter's rulers that were simply folded around the package and that would unfold under their own once released. Simple. And again, passive. We were concerned that the less-than-ideal conductivity of the steel might cause signal losses, but experimentation showed that it was insignificant. Turner Industries in Melbourne made hand tools for wood and metal working, and steel tapes. I still have a piece of steel tape that was cut off from the end of one of the antennas during the tuning process. And I still wonder at the piece at the other side of the cut that is still wandering daily around the Earth with its feet, inches and centimetres marked on it.

Tuning the antennas was another sign of the times. Almost the entire crew trooped down to Les' home in Clayton, Melbourne, where he had the equipment to tune the antennas. It was a beautifully sunny day with Les' wife sunning herself on a sun lounge. Concentrate on the job, boys!

Les had a radio signal strength meter, which we used to maximise the signal but, of course, it was changed by the presence of anyone nearby reading the meter. Never mind, Les brought out his 22-gauge rifle with its telescopic sight so that we could read the meter some distance from it. As if to prove this, I have a picture of me looking through the sight at the meter[2].

BATTERY

The Australis batteries were an alkaline manganese chemistry chosen for their slow self-discharge rate and relatively high capacity. They were a new type of battery then, with the patent only having been granted in 1960. The chemistry is the same as the modern 'alkaline' battery available in supermarkets and stores everywhere. However, each cell in Australis was considerably larger and enclosed in a cylindrical steel case.

The terminal voltage of alkaline cells is 1.5 volts and so fourteen cells wired in series made a single 21-volt string. There were two such strings, each supplying one of the two transmitters. The remainder

---

2   For disbelieving American readers, guns are very tightly regulated in Australia following the massacre of 35 people and 23 wounded in Port Arthur, Tasmania, in 1996. It would be unthinkable in Australia for a rifle to be used in this way today. There will be no discussion about the pros and cons of gun control in this book!

of the electronics—that is, the command receiver and decoder, telemetry system and HI Keyer—were supplied by both strings via isolating diodes. In this way, we were assured that one transmitter, telemetry and command system would continue to operate even if one battery string failed. Should one string fail due to an excessively high current drain in one of the electronics systems, for example, the other string would continue to supply its electronics, thereby giving Australis some redundancy. Individual cells were encased in epoxy resin.

Union Carbide supplied the batteries, but due to the long time between construction and launch, the batteries were replaced in the US prior to launch.

POWER SWITCH

Obviously, Australis had to be dormant and not drain its batteries until it reached orbit and was ejected from the launch rocket. Two micro switches mounted on the side of the structure closest to the launch rocket applied power when Australis was released from captivity.

STRUCTURE

In many ways, the structure was the most important single item of Australis as every sub-system required its support. Paul's superb manufacture of the structure amazed me with its perfection – my very amateurish attempts at woodwork were well and truly put to shame by his craftsmanship and accuracy. The electronics captured attention but the less glamorous structure (and stabilisation and orientation sub-systems) were equally important and significant.

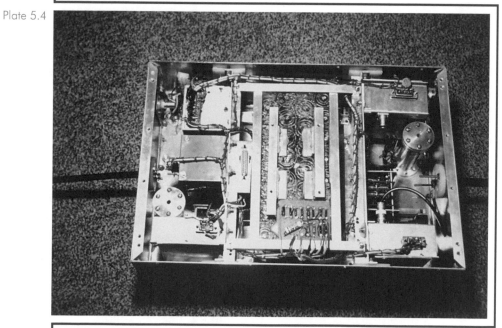

Plate 5.4

The shape of Australis was determined by OSCAR 3 launched in March 1965. We asked the OSCAR people whether that was the right shape and dimensions for our satellite and they replied, 'close enough'.

Plate 5.4 *View from above of the almost completely assembled Australis. The VHF antennas are installed (left and right ends), the enclosures for the two springs are clearly visible and the teflon thermal isolation between the internals and the outer skin is visible on the panel at the top of the picture.*

The structure was built around the heaviest and largest item: the battery. Electronics boxes were mounted on either side—HI keyer, HF and VHF transmitters on one side and Telemetry and Command Receiver and Decoder on the other. Magnets and hysteresis rods lay on the long side but as far apart as possible—the magnet was attached to the baseplate while the rods were glued to the top sheet of the structure. Antennas were mounted on Teflon pads on each of four sides, horizon sensors viewed space through small holes in each of the defined primary axes of Australis, that is x, y and z axes.

A layer of Teflon was used to thermally isolate the battery and electronics from the outer case, which became a thermal shield. The launch authority designed and applied a thermal control coating to the outer case, in this case a partial copper coating.

Other than the Teflon thermal and electrical isolation, the entire structure was made from an aluminium alloy (1.0 per cent magnesium, 0.6 per cent silicon, 0.2 per cent copper and 0.2 per cent chromium).

Rod wrote:

My contributions to the satellite were the antennas and the four mounting bolts or locator pins. The satellite was located on the launch vehicle using four conical locator pins on the side of the satellite that mated with four matching indentations on the launch vehicle. The specifications for these pins were quite tight and the angle of the cone was critically important. This angle was required to be within one degree of the specification. I made these locator pins by turning an octagonal aluminium rod using a small lathe in my home workshop. The hardest part of the process was getting the angle of the cone exactly right. I did this using an accurate template for the angle and a low-power microscope to make accurate observations of any deviations between the template and the pins.

Following the lead of earlier Oscar satellites, we chose to use metal measuring tape for the antennas. For a first prototype, I bought a number of retractable Stanley tapes at the local hardware store, Carlton Hardware, on Lygon Street, Carlton, and removed the tape from the spring-loaded cases. I cut the tapes to length and epoxied them into slots in pieces of Teflon approximately 5mm thick. These Teflon insulators were attached to the outside of the satellite. For the final version of the antennas, we used tapes provided by Turner Industries.

Plate 6.1 *Preparation for a balloon flight using a weather balloon being filled with hydrogen by Richard on the roof of the Natural Philosophy building. The single helix antenna is in the background.*

CHAPTER 6
Publicity—How We Told Our Story

Somehow, even at boarding school, I was aware of the value of well-placed and timed publicity. In those days, large schools had army cadets corps and certainly my school had a large and active corps. Dressed in khaki uniforms each Tuesday afternoon, World War 1 Lee-Enfield 303 rifles were handed out and the corps paraded around with them before shooting at the practice range. I was always amazed at the longest range, which must have been several hundreds of metres long, at least; but, disappointingly, I never got to shoot one.

I was a bandsman, improbably tooting on an improbably large trombone. I did enjoy band practice and contributing to the cacophony of our military band. I did not enjoy marching with the band, but the event that was most difficult for me was the annual cadet camp at Puckapunyal army base, north of Melbourne. Being an asthmatic (and at times it was very severe), the idea of sleeping on a 'palliase'—a large hessian bag filled with straw designed perfectly and specifically to be uncomfortable and to stir up asthma—did not appeal to me.

As senior boys were able to elect to do community service in place of army corps, I elected to do so in my final year at school. However, the band leader worked on me and finally persuaded me to continue with the band. I agreed, on condition that I did not have to go to camp. When the time came for the camp, I discovered that the master in charge of the corps had not heard of the arrangement and I had to visit Puckapunyal for the last time. I was not a happy bandsman.

The moment for revenge came when I read an article by a certain Lieutenant Colonel Flood, who was controller of army catering. He had defended army food and, as it happened, I agreed that the food we were provided with at camp was good. I wrote, with impeccable timing, to the afternoon newspaper complimenting the army cooks and staff. I wrote to emphasise the matter, after identifying the school: 'The school has a large and modern kitchen with staff proportionally three times as large as the army staff at Puckapunyal. Even so, the army food is far better than school food.' And I went on to congratulate army cooks.

I had timed the letter to be printed during the final school holiday and two weeks before returning for my final term. To emphasise the matter, the afternoon Herald newspaper ran a two-column article, with pictures, titled 'Army chow "up to hotel standard"' and with a photo of Sergeant DB Mace in his cadet uniform. (They got the initials wrong—Owen Bullivant Mace, the middle name being my mother's maiden name.)

Surprisingly, no letter arrived at home suggesting that I should not attend the school, nor was anything said on my return, although I did make sure that I was seen as little as possible by masters. Except for the French teacher, who rightly ejected me from his class during that term. Nevertheless, and just to spite him, I managed a pass in the French exam. I didn't know then that he was a heroic and exceptionally brave commando in the islands north of Australia during the war and, if I had, he would have received a good deal more respect from me.

I learned three things from this episode: the press can have a huge influence, timing is everything, and Jim Glover was a very modest war hero, as were others on staff.

* * *

During my first year as a member of MUAS, we had begun collecting weather satellite pictures for the Bureau of Meteorology, as has been described. There were occasional press releases and certainly the Bureau proudly used these new pictures from space in their forecasting, but with not a great acknowledgement of our efforts.

Once we had taken the decision to build a satellite, we realised that we had to become known in the community and so we began a publicity campaign. We were not total neophytes to the publicity game; after all, we had issued press releases through the student's union on a few occasions after listening

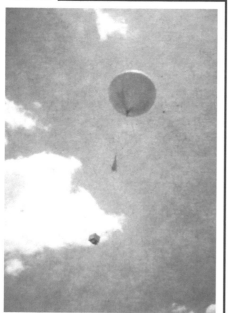

Plate 6.2

to various satellites. It seemed that simply issuing a press release on official University letterhead – there was no Public Relations Office then—would result in the press, radio, television and magazine reporters calling us. It should be recalled that space had captured the public's attention. President Kennedy's challenge to put a man on the moon by the end of the 1960s meant that preparations were well underway and various manned space flights were leading up to the flight of Apollo 11 and the moon landing on 12 July 1969. Interest in space was building with these reports, including photos and vision being broadcast onto television sets with almost every news program. The press were piqued by the thought that even in l'il ole Melbourne, there were people dabbling in space, and University students at that.

Our first Australis press release in April 1966 announced our intention of building a satellite and the newspapers responded better than we could have hoped. I have articles published by *The Sun* and *The Australian*. The front page of *The Australian* on 15 April carried a picture of four of us clustered around a mock-up of what we envisaged Australis would look like—a striped box about 300mm square with sticks emerging from four sides. How young we looked and how optimistic we were in expecting Australis to be launched by the end of that year. Nevertheless, there we were, ready to conquer space for Australia.

Plate 6.2 *Launch and away! The parachute and payload are hanging below the balloon.*

We were not sufficiently sophisticated in the art of public relations to have a single person designated to speak on behalf of us all, so many did and the press sometimes wanted to interview a particular person, such as Paul, who had a couple of large articles that described his interest in amateur radio. Paul expressed his belief that the requirement for proficiency in Morse Code was outdated – it has since been removed from the amateur license test. He had a World War II aircraft radar that he could use to detect clouds at a distance of up to 160km, anticipating today's weather radars. He made an eloquent argument for communications satellites and video telephones, once more anticipating what we consider as normal today. *The Sun* recorded that Paul's ambition was to build automatic landing devices for planes, among other ideas.

The Sun carried a brief article noting that we had 'a devil of a time thinking of the right name'. Apparently, we rejected 'Boomerang' in favour of Australis on the grounds that it 'would be embarrassing if it decided to come back'. The article reminded me that a large number of names for the satellite were canvassed. Australis seemed to me to be a natural name for an Australian satellite.

There followed discussions about the strict accuracy of reports and I argued that it was not as important as getting our names and faces into the press. My argument was that when people read a newspaper or magazine article, hear a radio broadcast or see a television news program, it is only the broad facts that are remembered, not the technical detail. We should, therefore, aim for our activities and plans to be published frequently and in all mediums.

Although I kept copies of all the newspaper and magazine articles that I found, I see that there are none concerning our collection of weather pictures. I have a dozen or more slides of those pictures, however.

* * *

By mid-year 1966, we had a clear idea of what we wanted to achieve, and the name we had chosen, and so now was the time for a publicity blitz. *Electronics* magazine described Australis as being about 430x300x150mm and weighing around 20kg. The August edition of the same magazine named the OSCAR team in California that included two we would subsequently meet, Lance Ginner[1] and Harley Gabrielson. It also described the progress of arrangements for the launch and reception of Australis' signals. The daily newspapers also carried articles, some multiple columns with pictures.

Plate 6.3

1   See http://www.arrl.org/files/file/Technology/Bilsing.pdf for pictures of the young Lance. Accessed 31 July 2017.

Plate 6.3 *Our simple payload landed a hundred kilometres east of Melbourne and was recovered. The remains of the balloon with its 'float valve' and my VW keys are visible.*

∗ ∗ ∗

Soon after our go-ahead came from Project OSCAR in 1965, it was decided that we should conduct a number of tests of the various sub-systems, especially the two transmitters. There were at least three reasons, including to publicise the project and to discipline ourselves to meeting a deadline. The third aspect was to exercise ourselves in the logistical requirements to achieve an end, which today is called testing our project management abilities. All these three goals were seen as important and, in reality, they were a mature, professional judgement of what is required for success.

As the reader can imagine, operating a satellite sub-system in an environment similar to that expected in space is not straightforward. We needed a hard vacuum and a solar radiation (heat and light) environment similar to that in orbit. The Army Test Establishment at Maribyrnong in western Melbourne had thermal chambers for testing army equipment, which is used at ground level. It also had vacuum chambers for testing Royal Australian Air Force (RAAF) equipment. None of these could simulate the space environment that we were expecting. We needed to get into or at least close to space, but then how to do so?

Richard's relationship with the Bureau came into play. He had been employed to fly weather balloons for the Bureau. These balloons carried small packages that measured wind, air temperature, pressure and so on and were carried on small rubber balloons to altitudes approaching 40km. This would be adequate for our purposes and had the considerable advantage of being far more newsworthy than a vacuum chamber in an army establishment. Furthermore, Peter's professor, Vic Hopper, had been experimenting with a means of 'floating' weather balloons at a certain, fixed altitude. He achieved this with a simple and ingenious contraption consisting of an aluminium tube inserted into the filling tube at the bottom of the balloon. A spring held a ping-pong (table tennis) ball in place at the inner end of the tube. A string was attached to the ball and the top of the balloon so that when the balloon reached a certain size, the string lifted the ball off the tube, allowing gas to escape and thereby stabilising the altitude of the balloon. The key to the contraption working as hoped was the tension in the spring, the length of the string and the attachment of the string to the top of the balloon. All too often, these were not quite right and the balloon burst, as they did without Prof's tube. Professor Hopper had flown a number of balloons finding the correct combination of tension, length and attachment, and he was able to float balloons, but not with 100 per cent success, as the balloons were never designed to have a string attached to the apex where, it is true, the latex was thicker and, presumably, stronger. Further, the balloon material became brittle in the cold temperatures and frequently broke.

As an aside, Vic was an experimentalist interested in getting into and above the atmosphere. His research interests were gamma ray astronomy, which was conducted on very, very large balloons. More on that later. He flew cameras on tethered balloons over archaeological sites in Syria, thereby anticipating the use of drones for exactly the same purpose decades later. With his balloon and camera, he helped the Australian archaeological team discover features that were not visible from the ground. He proposed large tethered balloons with electricity generators hanging beneath them, anticipating a renewable energy source. My favourite exploit of his was a tethered balloon flown from the University. The tether line broke, but, much to his surprise, the line remained aloft for over thirty minutes before being draped for many hundreds of metres across Melbourne roads, parks and homes. Imagine what the startled motorists and pedestrians would have thought on seeing a string gently settling across their path! Our infant son was a favourite of his when I joined his department a few years later.

Nevertheless, weather balloons were our chosen means of testing Australis sub-systems and a number of flights were conducted from the roof of the Natural Philosophy building within the University of Melbourne campus, which is barely a kilometre from Melbourne's central business district. I have numerous slides taken on glorious sunny days while preparing for a flight. And one of me showing how strong I was carrying a 50kg steel cylinder of hydrogen gas from ground level

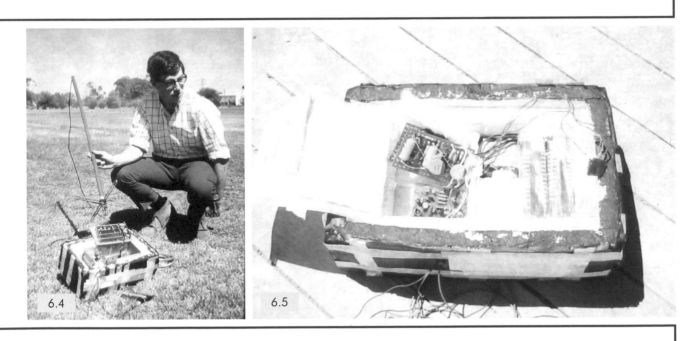

6.4

6.5

Plate 6.4 *A test translator package that was flown in February 1968. The radio frequency electronics is sitting on the polystyrene box and a 'baro-switch' to turn on the package above a pre-set altitude sits on the grass in front of the package. Richard is holding the 432 MHz transmit antenna. Amateurs' conversations, called contacts, were successfully conducted through the package.*
Plate 6.5 *A polystyrene package with a HF transmitter and HI keyer was flown in December 1967 on a HIBAL flight.*

up two flights of stairs to the roof. Trying to calm my heavy breathing and aching shoulder once on the roof, I decided that I would not be so stupid next time and ask for help.

There were discussions held on what to call these test flights and we settled upon 'Bravo' for balloon and on the basis that the 'B' should follow the 'A' for Australis, I suppose. In any event, three Bravo balloon flights were planned, beginning with the VHF transmitter and HI keyer. The flight was publicised by the Wireless Institute of Australia (WIA), whose Australian headquarters were located in Melbourne, convenient for us. Its signals were heard at Birchip in the Wimmera some hundreds of kilometres from Melbourne.

Perhaps the launch of one of these balloons should be described. Once the day had been decided, upon, it was always a weekend so that amateurs could listen and we students had no academic commitments. Peter arrives with his transmitter and test equipment, checks it out and gives the thumbs up 'okay' for flight. While the electronics are packed into the polystyrene box, the hydrogen gas cylinder is prepared with regulator and hose attached. Richard, experienced in flying weather balloons for the Bureau, begins inflation. Once it reaches about 2.5 metres high and 1.5m in diameter, the balloon is ready. Remove the filling hose and seal the balloon if it's one without a float device. Attach the radar reflector and, below it, parachute and payload, which is a mere kilogram or so.

Now remember that we are planning to launch this balloon from the University of Melbourne, which is close to the city of Melbourne and perhaps ten kilometres from Melbourne's airport, the Essendon Airport. Fortunately, we could see aircraft approaching the airport and so it occurred to us that some might not see quite the same excitement in our balloon flight as we did. Indeed, it might even be seen as a 'hazard to aviation'. Recognising this likelihood, we waited for the aircraft to pass before launching. I am corrected by Richard who claims a better memory than me, and quite possibly he is right. We telephoned the Civil Aviation Authority (CAA), as it was then, asking when might be an appropriate time to launch. One of the newspaper photographs even shows Richard with telephone in hand, no doubt deep in conversation with the CAA. As unlikely as such co-operation seems today, the okay was given and we duly launched at the suggested time.

By the time that we were ready to launch, the balloon, radar reflector, parachute and payload stood perhaps as much as ten metres above our heads. Needless to say, low winds were necessary for a successful launch.

The launch consisted of a theatrical count down and then simply let go of the payload. Richard followed the ascent using a Bureau theodolite until it disappeared in cloud or behind a building. Everyone rushed to our garret in the rooftops to listen to the HI signal and measure whatever we measured on the flight, mostly internal temperature.

If it was a clear day and everything went well, Richard might have been able to track the balloon for a considerable distance. Of course, the balloon expanded as it rose and the atmospheric pressure dropped, making his task easier. In truth, we had no way of accurately, or even inaccurately, knowing where the balloons took themselves, instead relying on farmers to find their remains and telephone the number written on the side of the payload. At least one flight was recovered this way from Avenel, north of Melbourne in Victoria, and another probably landed east of the city around Lilydale.

Plate 6.6

We were keen to recover our payloads, especially the electronics as there were valuable transistors to be reused. Cars, including my trusty VW and on one occasion my parent's Holden FB station wagon (complete with the fashionable fins of the day), set out with receivers and aerials to pinpoint the payload. I don't believe that we found any of the payloads this way, but it was fun racing around the Victorian countryside, stopping from time to time to listen to the signals while the balloon was still in the air. We knew the general direction of the balloon by using a directional aerial but 'pinpointing' would be somewhat an overstatement. A couple of cars could triangulate the position in theory but the beam width of the antennas limited the positioning accuracy. Nevertheless, it was clear the direction that the balloon took and when it burst and the payload was descending. Once on the ground with the aerial lying on the ground, signals were so weak that they were unreadable unless the receiver was very close to it.

Three flights were conducted; Bravo 1 carried a test two-metre VHF transmitter with HI keyer and some telemetry. Bravo 2 was similar and this time a ten-metre HF transmitter was flown. The third flight tested the command receiver, but by this time the CAA was becoming nervous about the risk of causing mayhem to aviation—this was to come some years later.

We were asked not to launch Bravo 3 from the University, so we moved to the Dandenongs, a range of hills—Melbournians call them mountains—to the east of Melbourne. We set off early one winter's morning to the gloriously named Mount Toolebwong to launch a flight to test the command system.

Despite everything, each flight was successful in that we managed to get all the equipment and people together and, furthermore, our electronics worked satisfactorily. Excellent.

Plate 6.6 *This is a high altitude HIBAL payload, similar to one on which we piggy backed, hanging from the 'hook' on the launch truck with the parachute in front of the truck. The balloon and parachute are tensioned ready for launch. Our package was taped to the main payload on the lower right side. The helium tank for cleaning the filter pads and an air intake funnel rising out of the payload are visible.*

Importantly also, the press followed us and gave us the publicity we needed and I began to record our activities with my half-frame slide camera. We could now press ahead with building the flight hardware.

Hard to believe these days with our digital cameras that allow us to snap away without concern for the number of pictures taken and that pictures, and movies, were recorded on film. There were two popular film formats. One approximately 7x10cm negative film from which prints were made at a size of choice. The other was a '35mm slide' where the image was recorded in a 35x23mm positive film. Both required the film, sealed in a lightproof canister, to be taken to a chemist (pharmacy) to be 'developed' and prints made, if required.

Following development, 35mm film images were mounted in 50x50mm cardboard 'slides' ready for showing with a slide projector that, between blown globes, threw the image onto the wall of a darkened room. The advantage of the more modern slide technology was that it was less costly and more likely to be in colour, but viewing required a projector.

With an eye to economy, I had a 'half-frame' slide camera that used half a standard 35mm slide but was mounted in the same cardboard "slide" and gave me twice the number of images per roll of film.

I often chose high speed 'Ektachome' film, which has not fared so well over the decades with the reds fading somewhat, giving a distinct blue tinge to the images. Resolution of half-frame is half that of full frame and there is evidence that some of the dyes of my Ektachrome images have flowed.

Nevertheless, with Paul's far more professional photographs, these are the majority of the images of those days. I have thousands of half- and full-frame slides from my late school days to the late 1990s when crude digital cameras began to make their appearance.

There were to be more flights, which, although they took place after Australis had been delivered to the US for launch, were another great adventure. Let's look at one now before moving on with the flight hardware and qualification testing. This is a little known story of nuclear espionage! There is very little recorded in the world wide web.

The independent minded French Gaullist government developed and tested a nuclear weapon in 1960. While early tests were conducted in the Sahara Desert in Algeria, the political instability of the former colony resulted in a relocation to French Polynesia in the Pacific in 1966, and many of the tests were carried out in the atmosphere.

The US wanted to understand the progress being made by the French and it chose to sample the clouds of very fine and radioactive debris, that is dust, from these tests as they spread around the globe. The composition of the dust debris would inform the US of the technical details of the bombs that the French were testing and their destructive size.

The dust rose to extremely high altitudes in the atmosphere, typically from 80,000 to 140,000 feet (24 to 42km). There was only one practical method of obtaining sufficient dust for analysis and that was by flying sampling equipment on very large, high altitude balloons. The balloon technology was well established by a number of companies, including Raven Industries in Sioux Falls, South Dakota (which I visited some years later), Winzen Research, Minneapolis, Minnesota and others. Using the cover of the National Science Balloon Facility (NSBF in Palestine Texas) and the US Air Force, the Central Intelligence Agency (CIA) regularly flew a series of flights at altitudes ranging from about 80,000 feet to 135,000 feet. Flights were conducted in South America, Southern Africa (I believe) and Australia. By the time the cloud reached Australia, it had travelled eastwards almost completely around the world and, presumably, dissipated somewhat, but even the dissipation might reveal something of the tests.

The large payload consisted of a pump that forced large volumes of high altitude ambient air through a series of fine filters, which collected the radioactive dust. The filter pads had to be recovered after the flight and returned to the US for analysis. Flaps and valves of the dust collection system were operated by a command system to ensure that the filter pads were not contaminated by dust on the ground and at lower than planned altitudes. There was what we thought of as an extensive telemetry system to measure air flow, altitude and other parameters of the flight. Position was measured surprisingly inaccurately by today's GPS standards, using the Loran C navigation system[2]. Once at altitude, valves and flaps were opened and the air pumps turned on for the few hours required to capture sufficient dust. Once the pumping was complete, valves closed, pumps turned off and the balloon "cut down" using the command system. A low frequency signal, below AM frequencies, was then turned on to track the payload down to Earth and allow recovery of the payload. Being such a low frequency, a very long antenna, some hundred metres, was released at balloon cut-down. The low frequency ensured that the signal could be heard even with the antenna lying on the ground.

---

2   Wikipedia, 'Loran-C' at <https://en.wikipedia.org/wiki/Loran-C>. Accessed 31 July 2017.

The launches of Project HIBAL, as the flights were called, were carried out from the Balloon Facility at Mildura, on the Victorian border with NSW where a Piper Aztec twin-engined aircraft was based. (I recall that the aircraft's registration was VHPOU.) The aircraft followed the balloon and its descent to the ground and then directed recovery crews to the payload. As a lead in to future chapters, scientists from around the world had realised that this was a relatively cheap means of lifting their instruments to near space.

The MUAS team also realised that this was an opportunity for flights longer than we could expect from a weather balloon, and probably higher altitude also. In December 1967, we flew an amateur band transponder from the HIBAL headquarters of the Australian Balloon Launch Service (ABLS) at Mildura airport. On this occasion, I took 35mm slide photographs, fortunately not with Ektachrome, recording the preparation, launch and recovered package. My photographs clearly show our package, as well as the dust sampling equipment. Even my VW features.

There were four further HIBAL flights in the first quarter of 1968 successfully testing translator electronies[3]

* * *

Let us now relive a typical balloon flight, such as we flew the test in December 1967, as they were to be part of my life for many years when I was engaged in balloon-borne infrared and gamma ray astronomy, noting that the atmosphere absorbs the infrared and gamma ray radiation that we wanted to measure. These are unmanned flights using very, very large balloons, some as much as 100 metres in diameter at altitude.

After a year, perhaps many more, the experiment is ready to be flown. That is, it has passed all the tests in our laboratory and has been calibrated, proving that it measures whatever it's supposed to and that it survives the pressure and temperature environment at altitude. The balloon has been delivered to the launch site and the launch crew at Mildura, Victoria, or the National Scientific Balloon Facility, NSBF Texas, USA, who have been contracted to fly our experiment.

The experiment and test equipment are packaged and shipped to the launch site, which is almost always a regional airfield – the places I've flown from include the NSBF headquarters at Palestine (pronounced 'palest – een'), Texas, and Malden, Missouri, in the USA. In Australia, the towns were Mildura, Broken Hill, Longreach and Alice Springs.

The time of year is carefully chosen to give as long a flight as possible to maximise observation time. The stratospheric winds at the altitudes that the balloons reach change direction twice a year, in March and September approximately, so flights are made when, hopefully, there is little or no wind at altitude.

---

3   Received on the two-metre amateur band and transmitted on the 432 MHZ Ultra High Frequency (UHF) amateur band.

Once again, the experiment is tested and then integrated with the command and telemetry package typically (but not always) provided by the launch crew. Once everything is ready, and having regard to the forecast weather conditions, a launch day is selected.

But first, we must understand something about the balloon itself. It is incredibly large and fragile. It is made from a specially formulated and thin plastic sheet, similar to the plastic that covers dry-cleaned garments after they have been cleaned. Strips of special plastic up to 100 metres long and two metres wide, called gores, are shaped and then heat welded with glass tape (for strength) in the seams. Top and bottom plates are attached to the tapes and, crucially, a rope passes through the bottom plate to one gore; its purpose will become clear shortly. Once complete, the balloon, now a very long bundle of gores, is folded carefully into a large box for shipping to the launch site. The boxes can be quite large, depending on the size of the balloon, up to six metres long, two metres wide, two metres high and weighing up to a tonne or even more.

Once inflated, the balloons are at the mercy of the slightest zephyr of wind and so dawn launches are much preferred as that is the time when the wind is at its lowest. Well before dawn, the equipment and balloon are taken out onto the airfield, and a long canvas sheet laid out downwind and carefully swept as the balloon will be laid out on this sheet. Pebbles, bindies and the like can easily puncture a balloon.

The balloon is now laid out along this sheet with its top end under a special roller. Once the experiment has been covered with insulation and the launch team's equipment attached, it is hung from the "pin", high on a mount at the back of the truck. The launch truck backs up to the bottom end of the balloon whereupon it is connected to the payload by the parachute. The long rope that is connected to a gore at the top of the balloon is tied to the parachute.

At the top end of the balloon, which can be as much as 100 metres away, helium gas is pumped into the 'bubble', which soon stands up high above the roller holding it down. More and more gas is poured into the bubble from huge ten-metre long cylinders on a semi-trailer until the carefully calculated and checked volume has been delivered.

To paint the scene, recall that the launch is timed to occur shortly after dawn, so the events so far have taken place in the dark with lights and torches darting, men in white overalls nervously marching around swiftly and the 'scientists' looking on, hoping for a good flight. As dawn breaks, the purples, reds and orange colours appear in the sky. Seen through the plastic of the 'bubble', dawn at a balloon launch is a memorable sight indeed.

Surprisingly, what is more memorable is the noise – let me paint the picture, if I can, of the launch, but first understand that the payload is held onto the launch truck by ropes that will be cut at the appropriate moment by explosive cutters, a .22 cartridge that drives the cutter.

As the launch approaches, more and more shouted, even walkie-talkied, commands are given: 'Check this, check that, is everyone ready? Yes.' Trucks that have been idling for hours are revved amid clouds of diesel smoke to ensure they are ready to accelerate rapidly. Finally, the launch director gives the word to launch.

Bang. No, a very loud metallic crash as the roller is released and the lift of the now-unleashed balloon rams the roller into a catcher.

The rustle of the balloon plastic overtakes other noises as the bubble drags the length of the balloon off the canvas and into the air.

Then the launch truck roars into life and takes off after the 'bubble', aiming to get underneath it before it passes over the airport boundary. The aim is to get the payload underneath the 'bubble' so that the payload is lifted off the truck and does not swing into the ground. But look, the truck is veering off to the right—the wind has changed a little!

As the truck disappears behind its dust cloud with the balloon standing majestically above it and the noises recede, suddenly there are two loud bangs and the payload rises.

A successful launch.

And now the work really begins monitoring the experiment, sending commands and getting excited by the results flowing in, but not before we've watched the balloon rise and recede into the distance.

Plate 6.7

Using binoculars, we can see that the helium gas is expanding and filling out the balloon. In time, it will reach float altitude once the gas has filled the balloon and excess flowed out its open bottom fitting. If we are lucky enough, we may be able to see it at float, looking more like a tiny upside-down onion than anything else, with the sun glinting off the plastic.

After some hours, as the balloon heads towards the horizon, the Piper Aztec plane VHPOU takes off to continue tracking. For whatever reason—batteries dying, getting close to the coast,

Plate 6.7 *Filling the 'bubble'. This is a relatively small (but still huge) balloon as there is only one filler tube. Note how the balloon is held under a roller on the trailer, and also note the distance to the parachute, launch truck and payload.*

whatever – eventually it is time for the flight to be ended and so the cut-down command is sent. While we cannot hear the explosive rope cutters that cut the rope between balloon and parachute, it separates them and the cord reaching to the top of the balloon rips out a gore, and the gas escapes, turning the balloon inside out. Remarkably perhaps, the balloon, now more like a candle, reaches the ground well before the payload on its parachute. Both are spotted by the plane, which guides the recovery crew.

That's the theory and mostly that's what happens but, as the reader might imagine, there can be incidents and disasters. Some of the more amusing incidents are recorded in Appendix D.

It was just this type of flight that we flew our December 1967 test from Mildura airport.

* * *

1967 was when our efforts to publicise the Australis project began to pay off. We began to target suppliers of the components we required. Most were very generous indeed and supplied us with whatever we asked for and they were acknowledged in the Users' Guide described later. The list is in alphabetical order:

- Acme Engineering, Melbourne, gave us radio frequency connectors.
- Cannon Electric Ltd, Melbourne, provided resistors and connectors. The resistors (electronic components) were high stability, high accuracy tin oxide components not then normally used in ground-based electronics.
- Ducon Condensers Pty Ltd, Sydney, supplied all the capacitors (also electronic components) used in Australis. Certain capacitors defined the audio frequencies we used in the command sub-system and so their temperature sensitivity was critical. They passed the test.
- Fairchild Australia Pty Ltd, Melbourne, provided all the transistors that were used, including very expensive space qualified transistors especially imported for us from the USA.
- Melbourne University (Student's) Union gave us a generous grant for ground equipment. This was probably the $400 of cash that I recall we had.
- Plessey Components Group in Sydney donated a travel grant, although I have no idea how this came about.
- The Potter Foundation in Melbourne gave us two travel grants. More on this wonderful incident later.
- Pye Pty Ltd, Melbourne, provided all the radio frequency crystals that set our two transmitted and command receiver frequencies.
- Rola Co (Aust) Pty Ltd in Melbourne supplied the magnets for the stabilisation sub-system.
- Sample Electronics, Melbourne, supplied circuit boards.
- Turner Industries Ltd, also in Melbourne, provided the satellite antennas.
- Union Carbide Australia Ltd, Melbourne, supplied flight and back-up cells for the battery packs.
- The Wireless Institute of Australia gave us a generous grant for running expenses.

There are some interesting observations to be made from this list. First, the majority of these suppliers were in Melbourne because it was the centre for manufacturing in Australia then. It could be said that Sydney had a strong electronics manufacturing capability with STC, for example, and that is correct, but it was only natural for us to approach Melbourne offices of companies wherever possible. While some represented US and UK companies, there is a strong representation of Australian companies. It's probably fair to say that few, if any, exist today, and that Australia has given up much of its manufacturing expertise to overseas companies. But enough lamentation of this well-known fact—for now.

We also acknowledged in the Users' Guide the help we received from the Meteorology Department of the University and the Bureau of Meteorology, which was, and remains, headquartered in Melbourne. The Glaciology Department of the University also unknowingly contributed low temperature testing chambers; that is, a chest freezer capable of reaching temperatures below domestic fridges. Perhaps it was wise not to acknowledge their unintentional contribution!

Almost without exception, companies that we approached for their contribution did so willingly and, importantly, helpfully. Clearly, our publicity had hit our intended targets, the decision makers in companies.

* * *

There were a couple of notable visits to companies to acquire the components we needed and the trip to Henderson Spring Works near the University illustrates that our efforts at publicity had been eminently successful. This factory made springs for cars, beds and many other uses but not yet for use in space. On appearing at reception, I explained who I was and that I wanted Henderson's to manufacture two matched springs to eject Australis from the launch rocket. Barely believing me, the receptionist asked me to wait while she sought someone to help, or

Plate 6.8

Plate 6.8 *Ready to launch! The launch crew are in white overalls and dignitaries in the foreground.*

hopefully to eject me. Evidently, the man who appeared at least knew of the project and he ushered me into his office, much to the incredulous looks of the receptionist. I explained what was wanted; such and such a size, so much force when compressed, able to withstand the unknown temperatures and so on. Realising that I understood at least something of what I was requesting, he agreed, telling me to return two weeks later when the springs would be ready for me to collect. They were. For me, it was a perfect example of how our publicity had smoothed the way for the project.

One other example is worthwhile telling. Although there was effectively no knowledge of what the radiation environment that Australis would find itself in, if only because we could only guess the orbit, it was reckoned that 'space qualified' semiconductors would be needed. By way of simple explanation, silicon transistors (not the field effect transistors of today) rely on charged electrons (and the absence of electrons) that move in the presence of an electric field in a chunk of silicon. The radiation that we were wary of consists of high energy ionising particles, some of which could penetrate well into the electronics of Australis and into transistors. The damage can be temporary, simply leaving a trail of charges that temporarily upset the transistor's operation, or it can permanently damage the very structure of the silicon crystal. 'Space qualified' transistors were designed and tested to be more resistant and resilient to both temporary and permanent radiation damage than conventional silicon transistors used elsewhere in domestic electronics, for example.

Plate 6.9

Several requests were met with refusals, not because they did not wish to help but simply because those companies did not manufacture space qualified transistors. However, we eventually contacted Fairchild Semiconductor in Melbourne who said that they would provide them for us. (Fairchild was a leader in solid state electronics at the time. However, it was not forward-looking enough for a few who later left to form Intel, the dominant manufacturer of the processor chips for personal computers.) Fairchild provided all the transistors used in Australis.

Richard and I made a trip to Sydney to ask Ducon Condensers for capacitors when, once again, my VW was invaluable. We set off for Sydney on Friday, 3 February 1967, at about 7:00am, hoping to reach Sydney that day. Incredible as it will seem to today's young, the Hume Highway had been only recently sealed. It was not uncommon in the 1950s

Plate 6.9 *Launch! What would Occupational Health and Safety people say today about the two crew on the back of the launch truck!*

Plate 6.10

for trucks to be bogged in the creeks and rivers that crossed the dirt portions of the road in New South Wales and for it to be impassable for many days in winter. Imagine the main road between Sydney and Melbourne being blocked! We expected a very long day.

As we travelled up Sydney Road past Brunswick and over Bell St, Coburg, we became aware of a large crowd of people on our right. 'What's this?' We asked ourselves. People seemed to be gathered outside Pentridge prison, and then we realised what the gathering was about. Ronald Ryan was about to be hanged. Ryan was convicted for the shooting and killing of a Pentridge warder, George Hodson, while he and fellow prisoner Peter Walker escaped from the gaol, hoping to flee to Brazil, which did not have an extradition treaty with Australia at the time. He was caught in Sydney three weeks later and extradited to Victoria, followed by his trial and sentencing. For the second time in my life I had a feeling of dread, this time over the imminent death of a man. I was unsure then of the rights and wrongs of capital punishment. (My view now is that it should be abolished, and it has been in Australia.) It was one of those events in life that I will always remember where I was at that moment, others around that time being the Cuban Crisis, President Kennedy's death and the Apollo 11 moon landing.

We drove on rather quietly as we contemplated what had just taken place. And on and on through Victoria and into New South Wales. And on and on through New South Wales until we were flagged down near Gundagai (but not '. . . nine miles from . . .'[4]) by a woman whose car had evidently broken down. As I brought the car to a stop behind hers, a couple of men appeared from behind a bush—this was not quite what it first appeared to be. She wanted fuel and the men seemed to be quite threatening, and so, fearful of what a refusal might bring, we allowed her to siphon fuel from the VW's tank. Unlike women of the time, she seemed to be adept at siphoning petrol, so it occurred to me that this was a regular gig for the three. Once she had enough petrol, there was a request for

---

4   This is a reference to a folk poem by Jack Moses: http://www.thedogonthetuckerbox.com/poemsfolk_songs Accessed 31 July 2017.

Plate 6.11

money, which we fulfilled, and drove away as quickly as we could. Thankfully, the rest of the journey was uneventful and Ducon gave us the capacitors we requested. I have a feeling that we returned via the coast.

We will delay the visit to the Potter Foundation until the appropriate moment in the story and follow the construction and testing of the satellite itself, the flight model.

Plate 6.10 *Release the payload! The truck manoevres under the 'bubble' so that the payload, when released from the launch truck, does not pendulum into the ground. The payload was released a few seconds later when the balloon had straightened.*

Plate 6.11 *HIBAL headquarters in Mildura in northern Victoria with my trusty VW.*

Plate 7.1 *Owen, Richard and Paul with the completed Australis satellite.*

CHAPTER 7
How We Built and Tested Australis

By late 1966, we had sufficient confidence in our ability to design and build our satellite and that it would successfully perform in space. Now was the time to build the flight sub-systems, assemble them and prove to ourselves that it performed as expected.

Flight components were being gathered together and delivered to the builders. Yet again, my VW Beetle proved to be invaluable as I journeyed between the homes of Paul, Peter, Dave, Les and Richard delivering components, receiving progress reports, listening to problems they had and goading them on. I saw Steve every day as we attended mostly the same lectures and practical classes, and we lived in the same residential college. As Australis progressed, the tempo of conversations with Project OSCAR increased in Les' radio shack. These journeys were not minor trips to the next suburb as Paul and Les lived some distance from Melbourne University and the college where I lived, so this became a not insignificant drain on my evening study time during my ever more difficult studies—third-year electronics engineering. It was by no means all lost time for me as I was visiting people who practised electronics by building circuits and using them. I learned a good deal about the practical side of electronics while studying the increasingly complex theory.

By the middle of third year, I was becoming increasingly anxious, even fearful, of the exams at the end of the year and even more so of the final year exams in November of the next year, 1968. I needn't have been but, nevertheless, it drove me to study even harder the difficult subjects of Mathematics—vector calculus, theory of variations and the like—and Physics—Maxwell's equations, relativity and so on. I enjoyed the electronics, especially, and remember my mother's concern when, in practical classes, we started to work with large electric motors and lethal voltages. My mother was certain that I would electrocute myself or become entangled in a rotating electric motor. Her concern was not without merit, but that was the point, to learn to understand the dangers and to protect oneself against them. If not, the same academic lessons could have been learned with low voltage, low power motors, etc.

Nevertheless, the Australis show had to go on as the sub-systems were completed, tested and assembled onto the Australis structure. Paul did a magnificent and truly professional job with the aluminium frame and structure. By about mid-year 1967, Australis was complete and ready to be tested as a whole. One issue that concerned us from the beginning was the manner in which the transmitters interacted (if at all) but certainly how they affected the command receiver. We needed to ensure that everything worked correctly and satisfactorily over the temperature range that we set ourselves. But first, the antennas had to be tuned and I have already mentioned the episode with the .22 rifle at Les' home.

I must pause in the story to describe our living arrangements. Already, we know that Steve and I lived in a residential college at the University of Melbourne. Richard shared a house near the University with Ian and a gentle and genial man who had nothing to do with Australis but who went on to a career with the Bureau. Crucially, however, three very eligible girls, Pauline, Delia and Ronnie, lived in a shared house on the other side of the University. May I say that Richard and Delia were friends? (The plot thickens considerably in due course, but, dear reader, you will have to wait a while for that to be revealed.)

Apparently, Richard had mentioned to Delia that he would like Pauline to 'come around and cook dinner for us one night'. History has not related why he would do that but in any event Pauline offered to cook a meal for Richard's household, a generous offer indeed. As recorded in an interview for a documentary on Australis, Richard had said that they didn't cook much and didn't eat very well. As Pauline remembered, 'the usual sob story, so I said I'll come around . . . to cook a roast'. She had already cooked the vegetables before she left her home, the roast was ready to go into the oven, but when she looked in the oven she found 'bits of metal and then this person materialises behind me with Buddy Holly glasses and he says "don't touch that, that's my satellite"'. Richard still wears Buddy Holly glasses. History does, however, relate that dinner was served about midnight, but only after Richard had completed the temperature tests on components of the satellite. It still amused me to think of the scene, even though, thankfully, I wasn't there.

Despite the initial meeting, the relationship between Richard and Pauline developed as we shall see. Pauline was and is a talented musician, then working as a violinist for the Melbourne Symphony Orchestra.

Now for the temperature tests on the whole satellite. It is true that the army test centre at Maribyrnong could readily undertake the temperature tests, but enquiries showed that they were unavailable at the time we needed them, so we had to be rather more inventive. I rather think that the temperature tests in Richard's oven that night sufficed for all high temperature testing of Australis.

Low temperature testing turned out to be somewhat easier. Our friend Ian worked in the Department of Glaciology in the University. Departments of Glaciology have freezers that can indeed hold satellites the size of Australis and so one weekend with Ian's help, Australis was fitted with temperature sensors, wrapped in plastic to exclude moisture (we hoped) and set in the freezer. I have pictures of Australis without its outer skins sitting in the freezer, wires attached and surrounded in mist.

Ian continued as a glaciologist, eventually working for the Antarctic Division until retirement. I met him a few years ago when I was given a flight to the Antarctic as a birthday present. It was Ian who was the resident expert on the flight and we renewed our acquaintance on board. Incidentally, the

12-hour flight was a thoroughly enjoyable experience with passengers and crew relaxed and enjoying the experience. We were invited to the cockpit, where I was even asked where I wanted the plane to go. My answer was appropriately Mount Melbourne, an active volcano leisurely puffing smoke into the atmosphere. The pilot pointed to a knob on the dashboard and motioned me to turn it to point the aircraft toward Mount Melbourne. The first and only time I have flown a Boeing 747 aircraft with hundreds of people on board.

All our testing showed to our delight that Australis would survive and operate as we had hoped in space. Frequency drifts with temperature were acceptable; we'd used transistors that were space qualified, so really, what could go wrong?

* * *

Around this time, David was experimenting with the University's IBM 7044 computer. His idea was that he might be able to use the computer's 'sense lines' to send and receive teleprinter characters between his desk in RAAF Physics and the computer, thereby being able to program it from his desk. An idea before its time. He achieved his goal and I remember him saying that there were several volts of 50 Hz AC mains difference between the computer and his desk. This required some additional electronics to extract the sense line signals from the 50 Hz, but readily done.

This meant that he could transmit results directly to his desk without having to wait to collect a printout. Mmmmm, this could be useful one day.

* * *

Plate 7.2

There is a moment in that year that particularly stands out for me. As Australis was nearing completion, it had to be delivered to the OSCAR people in San Francisco and this could be readily achieved by sending it via air. However, we thought, wouldn't it be fun to deliver it ourselves? Once again, we were unconstrained by what was possible, or rather not possible, and so we went about doing it. The question was how to raise the $2,000 per person for the flight. Imagine,

Plate 7.2 *Cold testing in the freezer at the Glaciology Department, University of Melbourne.*

Plate 7.3

THE AGE, Thursday, June 1, 1967

OUR FIRST SATELLITE

Australia's first satellite was being prepared for shipment to America yesterday, at the University of Melbourne.

From the outside, it looked like nothing more than a plain aluminium box, but as one of the m a k e r s explained: "There's 15 months of hard work in that and some of the most intricate radio equipment you'll find anywhere."

The satellite, named Australis Oscar is entirely an amateur project; it was built by members of the University's Astronautical Society and Radio Club.

It will be flown to San Francisco today where members will make final tests and make possible modifications on the satellite before handing it over to NASA for launching.

Australis Oscar will ride as a "passenger" in a

form-fitting alcove in one of the big orbiting rockets. As it climbs into space, an explosive charge will be activated, and Australis Oscar will shoot free of the rocket about 500 miles above the earth.

The antenna (made from steel measuring tapes) will whip out, and the two radio transmitters will begin broadcasting on wavebands which can be picked up by radio amateurs.

The Americans, possibly for security reasons, can give no information about the launching, but it is expected to be within the next six months, from California.

Australis Oscar is testbed for systems that will be used in a later Australian satellite currently being developed by the team and by Mr. Les Jenkins, a CSIRO technical officer.

Three members of the project team (from left) Owen Mace, 21, an engineering student, Richard Tonkin, 24, a law student, and Paul Dunn, 22, an applied science student, with the satellite.

Most dra[...] in GAS h[...]

Lord Casey
back today

$2,000 to fly tourist class across the Pacific in 1967. It was an enormous price and utterly beyond the means of all but a very few, certainly not us or our parents. We needed an organisation to give us the money.

We had an introduction to the Ian Potter Foundation, a major philanthropic organisation founded a few years before in 1964 by financier and stockbroker, Sir Ian Potter, who was strongly influenced by Kenneth Myer, son of pre-eminent philanthropist, Sidney Myer[1], and founder of the Myer stores. The introduction was most probably arranged by Willson Hunter.

However the introduction was made, Richard and I visited the office of the chairman on a glorious sunny Melbourne day in late summer. The address, as I recall it, was 95 Collins Street, Melbourne, which turned out to be Melbourne's only skyscraper then—of fifteen stories or so. Up the lift we went and into the Foundation's reception area. Two rather scruffy University students probably quite inappropriately dressed standing in front of an impeccably dressed receptionist. In due course, we were ushered into the office of the chairman, who turned out to be quite a tall and imposing man, also impeccably dressed.

The view from his office was quite staggering for me, who had never been higher than a few stories in any Collins Street building. Before us stretched St Kilda Road, the brilliant emerald green of the Domain and the Botanic Gardens, the brilliant whiteness of the Shrine and Government House and the blue, blue sky. What a view! This was, of course, the conversation starter for us as we exclaimed the brilliance and beauty of the view laid out in front of us.

Until, that is, Owen observed that it was all truly beautiful, except for the then uncovered Flinders Street railyards at our feet below, all dark and dirty. The chairman came over to me, put his arm over my shoulder and said, 'Not to me—I built those rails when I was railway minister in the Victorian government in the 1930s.' This was Sir Robert Menzies, Prime Minister of Australia from 1939 to 1941 and 1949 to 1966, the longest serving of all Australian Prime Ministers, a giant in stature,

---

1   The friendship between Potter and Myer led to the founding of the Florey Institute of Neuroscience and Mental Health, a major global player in health sciences.

both physically and in reputation. I reckon I am one of the few people to have had the arm of Bob Menzies around me, though whether it was in friendship, understanding, or he just wanted to throttle me, I don't know. People either admired him and his policies, or hated them both; there was nothing in between.

I do not remember the immediate reaction to this encounter, either from Menzies, Richard or me; we, nevertheless, explained our case and were dismissed with a promise to look into our request. We were granted two tickets! Somehow, Plessey Components Group were persuaded to give a third for Paul. Perhaps Willson Hunter at work again for us. So, three of us were to deliver Australis personally to Project OSCAR in San Francisco. The story of that incredible trip is told in the next chapter.

In the meantime, we had to decide who was going to go. No doubt all wanted to go but there were only three tickets. There was generous universal agreement that Richard and I should be two but the choice of the third was less clear. At a lunchtime meeting, it was decided that it would be put to a vote and that I would contact those not at the meeting that day. It is not with any pleasure that I remember telephoning one team member, explaining what was happening and asking for his nomination for the third seat on the plane. He was less than complimentary about the process and, unfortunately, we have never seen him or heard from him again since that day. Should he read these words, we would be delighted if he would contact us.

It was decided that Richard, Paul and I would travel. Peter ruled himself out as he was too busy with his PhD studies, besides it was unlikely that his professor, Vic Hopper, would allow him to go. The dates for travel, departing 1 June 1967 and returning 17 June, covered one week of term break and a week of the second term of my third year. Could I afford the time? I asked the kindly engineering Professor Moorehouse if I could go. His reply surprised me—'Yes, of course, you will learn more in those two weeks than we can teach you in a year.' He was correct.

Plate 7.4

Plate 7.3 *The Age, 1 June 1967, announcing the departure of Australis for the US. Credit: Fairfax Syndication.*
Plate 7.4 *Tuning a VHF antenna in Les' garden.*

Plate 8.1 *Our Boeing 707 in Honolulu airport 2 June 1967 behind the Japan Airlines 707. Minimal airport security then!*

CHAPTER 8
Delivery

In this chapter, I will describe what was for me one of my life experiences, and there were to be two that year in 1967. Richard, Paul and I travelled to California ostensibly to deliver Australis to the Project OSCAR group.

Although the British de Havilland Comet[1] aircraft had ushered in the jet age for long-distance civilian travel in 1952, fatal accidents from the then poorly understood process of metal fatigue prevented it from becoming the aircraft of choice. The Boeing Commercial Airplanes company, benefiting from de Havilland's experience, as well as from its deliveries to the US Air Force, developed the 707, a four-engine jet that has become the model for almost every subsequent passenger aircraft, namely engines slung below and forward of the low, swept-back wings, flying between nine and 12km at a little under the speed of sound. The design of the aircraft has to be admired since, despite travelling near the speed of sound, any air flow that exceeds that speed will cause shock waves that add drag to the aircraft. Although the 707 first entered service in the late 1950s, it was not widely patronised on the trans-Pacific route between Australia and the US. This was to be our first international flight.

We set off from Melbourne on 1 June 1967, proudly presenting ourselves at the Pan American (Pan Am) counter at Melbourne's Essendon Airport with our equally proud parents behind us. With the relatively inefficient engines, the flight in our Boeing 707 Jet Clipper was to be long with frequent stops to refuel. Auckland, New Zealand, was our first stop; then, I believe, Fiji. Perhaps it was American Samoa. There were so few passengers on board that each of us had three seats on which to sleep across and plenty of rows to choose from, too. As we were approaching Hawaii we were offered Hawaiian cocktails complete with a purple orchid floating in the conical cocktail glass. Very James Bond! However, not being used to such luxuries, I thought it tasted as if the attractive hostesses had raided the aircraft's fuel tanks and mixed in a bit of brake fluid. Shaken not stirred, of course. Richard recalls sampling Drambuie, a whiskey-based liqueur, which he became very fond of during the long flight.

More fuel at Hawaii before the final leg to San Francisco where we were met by OSCAR people who must have been surprised by the youths they saw before them. I had turned 21 less than three months before. Nevertheless, we were not abandoned but were taken to our "billets", and what amazing billets they were.

Now, I should explain that I was thoroughly enamoured by American technology, as you will no doubt observe in what follows. For a young nerdy engineering undergraduate, this experience was scarcely believable. Driving from the airport to Palo Alto where we were to stay, we passed SLAC

---

1   Built in Hatfield, England, where I was to work for a short time some twenty years later.

in Menlo Park. SLAC, the Stanford Linear Accelerator Center[2] was, and remains, the world's longest linear accelerator where atomic and solid state physics, chemistry and biology are studied. Accelerating electrons and positrons up to energies of 50 billion electron volts (50 GeV), it has provided discoveries of three fundamental particles and been instrumental in four Nobel Prizes. Later, it was the meeting place of pioneers of the home computer and personal computer. Here we were, Australian University students driving past this almost holy place!

Paul and I were billeted at Bill Eitel's home in San Jose and, yes, Bill was an amateur radio operator. (We learned to say 'San hose-ay' and certainly not 'San joe-say' or 'San-joze'.) He possessed a remarkably large home, what to us was an enormously large number of cars and, as we meandered his garden, a large number of gophers, judging by the holes in his spreading lawns; I had thought gophers

2 See https://www6.slac.stanford.edu/

Plate 8.2

Plate 8.2 *A fleet of cars outside Bill Eitel's home. Our Ford Galaxy on the left and his red Corvette with Richard talking to Bill's son.*

were just in the imagination of comic writers, but obviously not. His cars were remarkable for us — not just a Cadillac and a motor home, but also a red Corvette! Once inside, Paul and I met his wife and were shown to our bedrooms. Mine overlooked the gophers. It turned out that his neighbour, invisible through the trees, was the home of Judy Garland.

What was truly remarkable was not so much his cellar that seemed to occupy the entire footprint of his large home, and more, but that it was filled with all manner of lathes, drill presses, mills and other machine tools, the sort of things you might find in a well-equipped machine shop. We were to learn later that he had developed a major patent in this workshop! It turned out that Bill was a co-founder of EIMAC high power vacuum tubes.[3] Bill Eitel and Bill McCullough[4] were working for a manufacturer of vacuum tubes when, in 1934, they formed their own company, which grew to 1,800 employees that produced 4,000 radar tubes *per day* for the US government during World War II. Following the successful bouncing of a radio signal off the moon using 'a giant EIMAC klystron'[5], EIMAC became the largest independent manufacturer of high-power vacuum tubes. The company had recently taken over another manufacturer of high-power tubes, Varian. (No anti-competition laws then.) We were to meet the two Bills in their shared office later.

What was the technology developed in his cellar? It was a method for reliably joining glass to metal so that vacuum tubes and klystrons could operate at higher temperatures and hence higher power. This was probably one of the factors that propelled EIMAC to its prominent position in the industry. In fact, EIMAC manufactured cathode ray tubes for televisions that were marketed under the once-famous RCA name. Paul was a radio amateur and, of course, knew of EIMAC tubes as they were the preferred final amplifying tube for amateur radio transmitters. Nowadays, many are replaced with transistors. Bill Eitel died in 1989, McCullough in 2001.

Importantly, describing Bill Eitel, 'more than any other single individual, he is responsible for the OSCAR program'[6].

How little I knew of the history of wireless technology, amateur radio and OSCAR. How ignorant I must have seemed to the OSCAR people we met as we were taken to places and to meet people of significance to the four OSCAR satellites. In fact, I have learned much of this history from researching this book.

---

3   History San Jose, 'The Perham Collection of Early Electronics' at <www.perhamcollection.historysanjose.org>. Accessed 31 July 2017.

4   In the web page cited in the previous footnote, he is referred to as Jack McCullough. I am sure we were introduced to the two 'Bills' as humour was made of each calling the other Bill, and further confused when someone entered and asked for Bill. Perhaps Jack came to be used to distinguish between them.

5   A klystron is a special type of vacuum tube used to generate large amounts of power at very high and microwave frequencies.

6   *World Radio* magazine, May 1989, quoted in: https://qrqchistory.wordpress.com/about/

Richard was billeted with Bill Hewlett—yes, another Bill, and the co-founder of the Hewlett Packard company, which in those days made premier electronic instruments. Bill and his wife lived in a mansion in Palo Alto, and they even had a maid. They made Richard very welcome.

The day after our arrival, we were escorted to Foothills College, the home of OSCAR, by Harley Gabrielson in his Mustang (Mustang!). I regret to say, I understood nothing of the College significance to the people of OSCAR, but I certainly ogled at his car.

We were presented with a Ford Galaxy for our use while we were in San Francisco. The fashion for cars with large fins had died and so our Galaxy was 'fin-less', but enormous compared to my diminutive VW. I have a photo of me at the helm with the speedometer reading 90—and that is not kilometres per hour! The next day, we were sent off to drive to the natural wonders of the Bay Area, Sausalito, the Big Basin and Muir National Parks, north of the city. Being the only driver, it was my responsibility to get us there—my first experience with such a big car and with automatic transmission, too. To add to my confusion, the steering wheel was on the wrong side of the car

Plate 8.3 *Lombard Street, San Francisco.*
Plate 8.4 *Richard and the Galaxy (registration UBG 508) parked in Lombard Street.*

and—what's this?—the cars drive on the wrong side of the road as well! Richard, the navigator, was given instructions on how to get there via El Camino Real[7]. El Camino Real? Yes, we were told, it was a major road that would take us to the city of San Francisco and it had historic connections with the Spanish missionaries from the late 1600s. While Richard may have been 'the natural leader', perhaps navigation was not one of his strong points, so it was just as well that Paul had listened to the instructions.

First, a tour of the city of San Francisco and a drive down Lombard Street, one of the impossibly steep streets on Telegraph Hill, the one with eight tight hairpin bends. We parked the Galaxy below the bends to admire the city (plenty of parks in those days) and the well-planted street. With its soft suspension, the car leaned downhill as Richard appeared to lean toward it in my photograph.

To get to Muir National Park, we had to cross that icon of San Francisco, the Golden Gate Bridge, impossibly high over the waters and then on to Big Bend where we saw gigantic Redwood trees and tame deer. Another two firsts for us. Then Sausalito, where there were floating houses—another first. Back over the Golden Gate Bridge to visit Haight-Ashbury, the centre of the 1960s hippie movement[8], and what a sight it was for us innocent young Australians. We decided to extend our tour by circling the bay before returning to San Jose via El Camino Real with two more huge bridges and the notorious San Quentin prison *en route*.

On the next day, Monday 5 June, we were taken to the IBM computer plant with

7   Wikipedia, 'El Camino Real (California)' at <https://en.wikipedia.org/wiki/El_Camino_Real_(California)>. Accessed 31 July 2017.

8   Wikipedia, 'Haight Ashbury' at <https://en.wikipedia.org/wiki/Haight_Ashbury>. Accessed 31 July 2017.

Plates 8.5 *Famous names on a pedestal of the Stanford University Mills Cross radio astronomy antenna.*
Plates 8.6 *70 metre dish at Stanford's antenna farm.*

a sea of employees' cars in its car park. A memorable moment came for me when we were walking down a corridor and our guide stopped, looked around furtively, and then quickly ushered us into a rather stark room. It contained a desk with a rather large box, probably a metre cube, with a small, 300mm television screen visible on one side and some sort of keyboard beside it. This, we were told, was the future of computing. What we didn't realise then was that IBM was investigating multi-user, multi-tasking operating systems for its recently introduced IBM 360 line of mainframe computers[9]. These machines were truly pathetic imitations of computers when compared even to today's laptops. The machine on which I am writing this book is around a thousand times faster and carries a thousand times more memory. Nevertheless, these machines were pioneers in computer architecture. Furthermore, it is possible that the human pioneers of modern computing, IBM Chairman Thomas Watson and chief architect of the System/360 Gene Amdahl, were in the same building as us. Historic, but it takes the wisdom of remoteness in time to realise it as being so. Amdahl went on to found a company that made compatible, and less expensive, computers bearing his name.

Interestingly, specially modified and hardened IBM 360 computers, called 4 Pi avionics computers, were used as fault tolerant computers in the Space Shuttle. As a compliment to their advanced design, the Soviets made clones and several other manufacturers made compatible machines.

Plate 8.7

Tuesday saw the arrival of our pride and joy, Australis, so off we went to the airport to collect it and deliver it in its well-padded box to Lance Ginner's home. It had been delayed because of a last-minute problem with the command receiver. Nevertheless, here it was and it was opened on Lance's lawn – inside was a message, 'God Save the Queen'. How inappropriate and embarrassing now. Of course, republican sentiment in Australia has become much more popular since then. Oblivious of the future embarrassment, the three of us were photographed with the sign and Australis in its delivery box. Nevertheless, we had accomplished our mission to deliver Australis to Project OSCAR's representative, Lance Ginner. Little did we imagine then that it was to lie in Lance's garage for nearly two years.

The next day, Wednesday 7 June, we were taken to the Stanford University and the SLAC.

Plate 8.7 *Delivery in the US and the embarrassing message. Look at the fins on the car in the top left corner!*

9   Wikipedia, 'IBM System, 360' at <https://en.wikipedia.org/wiki/IBM_System/360>. Accessed 31 July 2017.

Thursday saw us visit the Stanford 'antenna farm', a large country area set aside for radio astronomy and antenna development. The largest antenna was a 60m diameter radio astronomy dish. There was even a US Air Force quad helix of similar dimensions, and therefore gain, as ours back in Melbourne. Copycats! I must say in its praise that it looked a good deal more substantial than ours and no doubt the large building next door to it contained a good deal more equipment than our meagre garret held, and a good deal more sophisticated. Next, we were shown the Mills Cross, an array of relatively small dish antennas perhaps two to three metres in diameter and arranged in a cross. The concrete pedestal on one was graffitied with the names of some intellectual giants who had visited before us—names like astrophysicist Tom Gold, Australian radio astronomers John Bolton, Bernard Mills and AG Little[10] (the latter two being inventors of the Mills Cross) among others. We were indeed in august company!

Lance was a young engineer, but older than us, with the Corona spy satellite project, though I knew little of it then. On Friday 9 June, we were invited to dinner with the insanely handsome, crew-cut, all-American Lance Ginner and his very attractive wife, Wanda, as this was to be our last day in San Francisco. Over dinner, it was announced that we were to go to our next set of meetings in Los Angeles. We had the weekend to drive and Yosemite National Park was a recommended destination.

Saturday morning saw us take possession of another Ford Galaxy. We were told that it was a hire car that had to be returned to Los Angeles, but I wonder whether it was not another example of their incredible generosity. This one took my breath away . . . and I was to drive it! It had all the features of our previous, yellow Galaxy, plus it was a convertible and it had the unimaginable—air conditioning. Air conditioning in a car! Furthermore, it was a screaming bright RED! Oh well, I suppose I just had to do my duty and drive it to Los Angeles.

10   See 'Early History of Radio Astronomy' at <http://www.astro. washington.edu/users/woody/hra. html>. Accessed 31 July 2017.

Plate 8.8

Plate 8.8 *OSCAR and Australis teams. From the left: Don Norgaard, Ed Hilton, Lance Ginner, Paul, Owen and Richard.*

Off we went, with instructions to present ourselves on Monday morning at. . . that will be revealed when we get there. We followed the now familiar El Camino Real past NASA Ames field with its huge dirigible hanger and eastwards to the Napa Valley and Yosemite National Park. While we were surprised at the number of people in the park compared with near zero in comparable Australian parks, it was nothing compared to today's annual four million visitors when it is necessary to reserve a visit!

Yosemite presented itself in all its truly staggering beauty. The iconic images we see of El Capitan, the redwood trees, the mountain meadows and passes, deer, valleys carved out of solid granite by ancient glaciers, Yosemite River in flood flow, or so it seemed to me. The effect that it had on me was in the number of pictures I took: forty-two in all and, unfortunately, all in Ektachrome, not yet colour-corrected in their digitised form. Remember that each photograph cost something, unlike today's digital photographs. Forty-two half-frame slides was approaching a full roll of film! Quite an expense for a young undergraduate.

After spending a full day at Yosemite and wishing we had more time to explore the high country, we set off toward our next destination, driving through towns with Spanish-sounding names like Merced, Madera and Fresno, arriving at the town of Lompoc, surrounded by kilometres of commercial flower farms.

8.9

8.10

8.11

Plate 8.9 *The Red Ford Galaxy at Yosemite National Park with Richard photographing Paul who had not seen snow before then.*
Plate 8.10 *Richard standing behind an engineering model of the lunar 'crasher', Ranger.*
Plate 8.11 *The NASA plane that flew us to the Goldstone tracking station.*

I am reminded of two incidents. The first occurred when we stopped to fill the car with petrol—no, it's 'gas' in the US. The attendant came to the driver's side—it was still the time that attendants pumped petrol, gas, and checked the oil—and said something. I knew it was a question by the upward inflexion at the end of his sentence but I understood not a word. 'Excuse me?' I said, and his question was repeated. Now we were visitors and it would have been rude to have asked him to repeat the question for a second time, so I assumed he was asking if I wanted the gas tank to be filled. I nodded and it was.

The second incident occurred as we were driving in the dark toward our overnight stop. Red and blue flashing lights behind us. Uh oh… I wasn't exceeding the 55 miles per hour speed limit so it couldn't be that. Once stopped, the impressive policeman sidled up to the car and asked for my driver's license and I presented my newly minted Victorian license and an International Driver's license, then a necessary accompaniment for foreign drivers.

'Where y'all from?' he asked.

'Victoria, Australia,' I replied.

'Ah, Victoria, Canada?'

'No, Victoria, Australia,' with the emphasis on Australia.

He got it this time and repeated, 'Ar-stria'.

Having established where we were from, he very politely informed me that one of our tail lights was not operating, 'Sir', and we couldn't proceed. Now this presented several problems, the first being a 21-year-old being addressed by a much larger, older and armed policemen as 'Sir' and the second being how to get it repaired at night and in a foreign country. He agreed that we would inspect the recalcitrant light and decide what to do. We were foreigners after all. So, I got out, lifted the boot, pushed the light fixture back into place, whereupon it lit up satisfactorily and we continued on our way with our newly minted policeman friend wishing us a safe trip back to 'Ar-stria'.

Richard recalls a third incident. We stopped at a cafe (a diner) in Lonpoc for lunch and he asked the waitress for a chocolate milkshake.

'Excuse me?' she asked.

Richard repeated the order, 'A chocolate milkshake, please.'

'Are you folks from out of town?' she asked.

A fellow customer came to our aid. 'Honey,' he said to our waitress. 'These folks sound like they come from England. The guy wants a chocolate malted, but hold the malt.' As Churchill, the World War II British Prime Minister, said, 'We are separated by a common language!'

Then on Monday morning, we presented ourselves, as instructed, at the gate of the Vandenberg Air Force Base. Now for us space cadets (as Richard is wont to refer to MUAS members), 'Vandenberg', as we knew it, was another near-holy place. Its mission was not to fly aeroplanes but to fly rockets and to launch satellites into orbit for the US military. It was from here that the first three OSCAR satellites were launched[11].

We announced ourselves to the burly uniform that arrived to escort us into the base explaining where we came from. 'Yesirrrr, Ar-stria,' came the reply. Oh, never mind. 'Leave your cameras in the car,' he commanded. That day, we climbed over a launch site for Scout rockets and, I think, the control room. No doubt that thanks are due to our OSCAR friends for the VIP tour of the base.

On to our destination, Los Angeles, where further amazing techie sights awaited us. We were very surprised to see oil wells in the shallows and pumps in gardens idly pumping that oil to refineries that were probably nearby.

On Tuesday 13 June, we presented ourselves (and our showy red Ford Galaxy convertible with its air conditioning) at the Jet Propulsion Laboratory,[12] another near-holy site for us. JPL, as it is universally known, has had a very long history in rocket development beginning in the 1930s as the Guggenheim Aeronautical Laboratory. During the war years, it was operated by the California Institute of Technology for the US Army and nowadays for NASA. Famous names, such as Theodore von Kármán and Werhner von Braun, and famous rockets are associated with JPL, including America's first satellite, Explorer 1. More recently, many spacecraft, especially lunar and planetary spacecraft, have been developed at JPL.

We were met by the likeable Ken Hornbrokk, who showed us around their hall containing models of legendary spacecraft, Surveyors, Rangers and Lunar

Plate 8.12

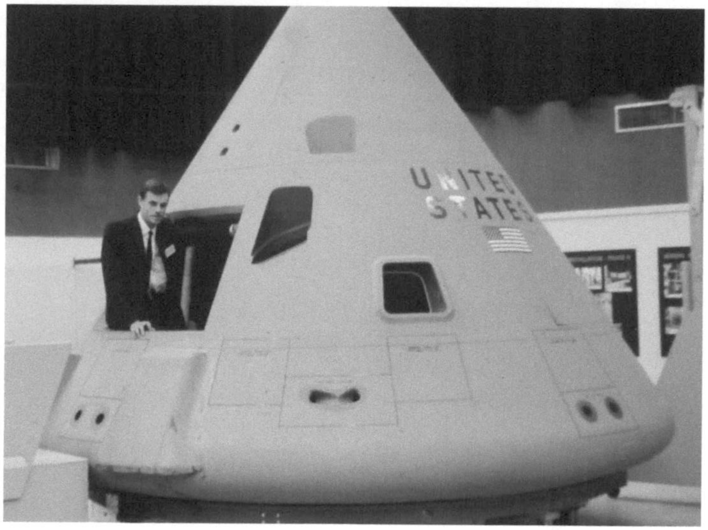

Plate 8.12 *Astronaut Paul checking out the Apollo Command Module for size.*

---

11   The fourth OSCAR, intended for a geostationary orbit, was launched from Cape Canaveral, Florida.

12   Wikipedia, 'Jet Propulsion Laboratory' at <https://en.wikipedia.org/wiki/Jet_Propulsion_Laboratory>. Accessed 31 July 2017.

Orbiter, displayed in front of us. In another room, the lunar Orbiter spacecraft[13] was being prepared for flight. We were in the presence of a real spacecraft, one that would be sent to orbit the moon in preparation for the Apollo landings!

There followed on Wednesday 14 June a visit to North American Aviation,[14] which later became part of Rockwell International and then Boeing. It was founded in the late 1920s as a company trading interests in the early airlines and in the 1930s it began to manufacture training aircraft. During World War II, it produced the Mustang fighter and the Mitchell bomber, among many others and later such iconic aircraft as the Sabre and Super Sabre of the Korean War as well as the Mach Three[15] Valkyrie aircraft[16]. Interestingly, other divisions of the company researched and built nuclear reactors and defence electronics. It was also heavily involved in the manned space program, which was what we were there to see. North America had won contracts to build the Command Module, the Service Module and the second stage of the enormous Saturn V rocket that carried the Apollo program to the moon.

Recall that a few months before our visit, the disastrous Apollo 1 fire killed three astronauts, adding yet further pressure to President Kennedy's December 1969 deadline to land men on the moon. Nevertheless, we were lead into a gigantic hall where Apollo command modules were being built. On the left, with assembly having been recently begun, was the Command Module for the first manned flight, Apollo 7, that was launched on 11 October 1968. Following a walk through the control room used to test the Command Modules, we were taken to an engineering mock-up of a Command Module and each of us invited in turn to climb into it. Each of us had our picture taken entering it.

In the midst of this excitement, we were taken to a place of exotica – an Armenian restaurant. Armenian? I had barely heard of the country and certainly had no idea of where Armenia was. This was a time in Australia of 'meat and three veg' for dinner every night and exotic alternative menus such as Italian pastas were just starting to be seen in Melbourne. This was a very different experience for us.

While researching the Apollo project, I came across this quote on the Wiki page[17] under the heading of Science and Engineering Legacy:

13   Wikipedia, 'Lunar Orbiter Program' at <https://en.wikipedia.org/wiki/Lunar_Orbiter_program>. Accessed 31 July 2017.

14   Wikipedia, 'North American Aviation' at <https://en.wikipedia.org/wiki/North_American_Aviation>. Accessed 31 July 2017.

15   That is, three times the speed of sound, around 3,000km per hour!

16   Wikipedia, 'North American XB-70 Valkyrie' at <https://en.wikipedia.org/wiki/North_American_XB-70_Valkyrie>. Accessed 31 July 2017.

17   Wiki, 'Apollo program' at <https://www.en.wiki.org/Apollo_program>. Accessed 31 July 2017.

The Apollo program has been called the greatest technological achievement in human history . . . The flight computer design used in both the Lunar and Command Modules was, along with the Polaris and Minuteman missile systems, the driving force behind early research into integrated circuits (IC). By 1963, Apollo was using 60 percent of the United States' production of ICs...

Given our interest in electronics (and later ICs) and technology, as well as their application in space, of course, we were privileged indeed to have seen what we did, even if we did not fully appreciate the full extent of our privilege.

We had been instructed to return to JPL that evening, 14 June, to view the control room during the launch of the Mariner 5 Venus fly-buy. This spacecraft was the back-up for Mariner 4, which was launched in November 1964 and flew past Mars obtaining the first close-up pictures of the Martian surface. Following the success of Mariner 4, the back-up was fitted with additional thermal insulation because Venus is closer to the sun than the Earth, whereas Mars is further away. Also, the size of the solar panels was reduced (more solar radiation than on a Mars flight) and a TV camera removed, presumably because the Venusian surface is covered by opaque clouds. Mariner 5 flew past Venus on 19 October 1967, successfully completing its mission and finding that the Venusian surface and atmosphere were hotter than expected.

With all the expected drama, the countdown began and proceeded second by second with the questions and answers:

'Atlas First Stage?'

'Go!'

'Agena Second Stage?'

'Go!'

'Mariner?'

'Go!'

'Communications?'

'Go!'

and so on through all the systems. Then, 'ten, nine, eight' and so on, interrupted by 'Cooling on' and 'first stage ignition', then 'Zero' and 'Liftoff!' The Atlas Agena and Mariner had begun their journey to Venus and yet we heard not a single roar of rocket exhaust, nor could we look out a window to see the night lit by the rocket's exhaust tail. What was unusual about this launch, besides being our first, was that it was the first to be conducted remotely. The launch took place from Cape Canaveral,

Florida, whereas the control room was in Los Angeles on the other side of the country. What a privilege to witness this particular launch!

The next day, Thursday, saw us arriving at yet another airfield where we were flown in a NASA Beechcraft twin-engined aircraft to the Goldstone Tracking Station in the Mojave Desert. More antennas, including the 70m Goldstone "Mars" dish used for tracking interplanetary spacecraft like the one that was launched to Venus on the previous evening.

There were two visits on our final day, neither of which permitted cameras. The first was to the Lockheed Corporation where, incredibly, we were shown a room where glass fibre was being wound around a very large mandrel over one metre in diameter. This, it was explained, was the rocket body for Polaris, a submarine launched intermediate range ballistic missile – nuclear warhead, of course. Did they really know who we were? Some highly important people cleared to the highest levels of national security clearances? No, just a bunch of foreign undergraduate students. I am sure we were shown other things that Lockheed staff were working on but I was just incredulous at seeing the Polaris rocket bodies in production.

The other, Hughes Electronics, produced all manner of radio frequency electronic devices for the military, space and other high-end applications. They were experimenting with stabilising satellites by spinning them and we were amused by one that seemed to be wobbling like a drunk. Next, we were led to a hall where Surveyor 4 was in its final stages of preparation for launch. Its purpose was to measure the characteristics of the lunar terrain in preparation for the later manned moon landings. What a thrill for us space junkies! In his enthusiasm, Richard asked if he could actually touch it. Yes, but only with a gloved hand.

Launched a month later, Surveyor 4 travelled flawlessly to the moon, and yet, some 150 seconds before it was to touchdown on the moon, its radio signals were lost and never re-established. It was thought that the rocket slowing it for

Plate 8.13

Plate 8.13 *Astronaut Owen about to depart for the moon!*

Plate 8.14

Plate 8.14 *Owen and Paul in front of a model of the Apollo Service Module.*

landing exploded and the mission itself was declared a failure. In later life, Richard wondered if it was his touch that caused the spacecraft to fail to complete its mission. Unlikely, but on the other hand, Richard...

This was the end of our trip. We had seen some of the most advanced technological achievements of the day and, in the Project Apollo, some of the most incredible achievements in management and organisation. For President Kennedy to be able to inspire and galvanise a country such as the US into such a sustained and challenging project can only be admired and treated with wonderment. As Kennedy had said when he spoke at Rice University on 12 September 1962:

But why, some say, the Moon? Why choose this as our goal? And they may well ask, why climb the highest mountain? Why, 35 years ago, fly the Atlantic? ...

We choose to go to the Moon. We choose to go to the Moon in this decade and do the other things, not because they are easy, but because they are hard; because that goal will serve to organize and measure the best of our energies and skills; because that challenge is one that we are willing to accept, one we are unwilling to postpone, and one we intend to win...

Inspiring words of the kind we need today to lift us out of mediocrity and scepticism.

When I look back at what we were shown, things that most Americans would never see and, in many cases, would not even be allowed to see, it is clear that we were treated as very special visitors. The generosity was simply staggering. The arrangements had the imprimatur of some important people. Willson Hunter must have had a hand in it. We will never know.

Departing Los Angeles on Saturday 16 June 1967, our trip home took us via Hawaii, Samoa, Auckland and Melbourne. I had missed a week of the second term but, as my Professor had said, I had learned more about engineering, technology and life from this trip than they could teach me in a year. The sad part is that there were but three places on the trip, but many more who would have learned at least as much as me. Now back to our studies – and the excitement of the Apollo flights culminating in the landing on 20 July 1969, just over two years later.

Plate 9.1 *Try launching three balloons at once! This is a test launch in 1979 of a "Super Pressure" balloon system that provided many-day flights. We flew one such balloon around the world in under two weeks, a record duration at the time.*

CHAPTER 9

Wait for Launch, or Would Australis Come Home in a Box?

Now back to our studies and our normal lives. Not quite.

The telephone rang a few weeks after our return at one of our usual lunchtime meetings. As I was nearest the phone, I answered it and on the other end of the line was a woman, one who sounded very, very embarrassed and most uncomfortable. She explained that she was from the Department of Supply, the logistics arm of the Australian armed services, and that she wanted to ask some questions about our satellite. After a few moments to gather my thoughts following such a strange announcement, I reasoned that we had nothing to hide (a jibe at the Department) and would be happy to answer any questions she had. The answers to her questions were nothing more than could be read in the paper, but they culminated in what was the clear intent of the call even at the time; when is Australis going to be launched? We understood it to be likely in December of that year, 1967, and so that's what I said. Thank you, goodbye. In the following discussion, we all thought it was very odd as there was nothing that could not be found by diligently reading the newspapers, which is what, after all, my spy father did for the Department of the Army, although his interest was Indonesia.

All was revealed the very next day when WRESAT,[1] a satellite from the Weapons Research establishment (WRE)[2], was announced. That was what the telephone call was about! This should not have been so surprising as I had written to our local member of Parliament, Peter Howson, Minister for Air, telling him of our intentions. I have his acknowledgement letter dated 4 March 1966, thanking me for my advice to the government.

In 1966 and 1967, the US had been conducting tests from the Woomera test range on the re-entry of nuclear warheads from space. Ten Redstone/Sparta rockets had been sent to Australia for the tests, along with a team to launch the rockets. Sparta was a liquid fuel Redstone first-stage rocket with solid fuel second and third stages. Redstone was developed by Werner von Braun as a modern version of the wartime V2 missile and it launched Explorer 1.

---

1   See http://www.honeysucklecreek.net/supply/WRESAT/main.html; and
     https://www.dst.defence.gov.au/innovation/wresat-%E2%80%94-weapons-research-establishment-satellite.

2   The Weapons Research Establishment (WRE) was a joint project between the UK and Australia for testing and proving military weapons, including rockets and missiles. (Its remit was widened and it became the Defence Science and Technology Organisation, DSTO, and now the Defence Science and Technology Group, DST Group.) The test range was, and is, located in the desert in South Australia and extends into the Northern Territory and Western Australia. The research establishments are located throughout Australia, principally in South Australia.

The flights had proceeded with one failure and so the spare Sparta was not expected to be needed. Someone suggested that the unused rocket could be used to launch an Australian satellite and the idea took hold. Let's rephrase the last sentence with an eye to the conspiracy theorists and say that, in view of the possibility that Australis could be launched before any official government satellite and much to the government's embarrassment, the unused rocket could be used to remove that embarrassment.

Needless to say, there is no proof either way, but that is the stuff of conspiracy theories after all. Nevertheless, WRESAT was launched on 29 November 1967, conveniently before the expected month of the Australis launch. That a launch before the end of 1967 was part of the agreement with the US is left unsaid by those conspiracy theorists.

NASA had agreed to track WRESAT, no doubt Willson Hunter again, and the US paid for the TRW launch team while the people at WRE set to work. In under a year, a complex satellite was built, tested and flown. It extended the work that WRE had been doing in the upper atmosphere with the University of Adelaide, in particular the heat balance between the sun and Earth, a pre-cursor to today's climate modelling. More specifically, detectors measured ultraviolet radiation and X-rays in four wavelength bands to understand the distribution of oxygen and ozone in the upper atmosphere, as well as the physical processes taking place.

There were also goals, parallel to ours, of extending scientific, technological and management disciplines.

Generously, the Department of Supply invited us to the launch of WRESAT at Woomera in the South Australian desert. We had to get ourselves to Adelaide and we would be flown to Woomera for the launch. Richard, Peter and I made the overnight train trip, with me, the smallest of us, sleeping in the overhead luggage rack, but the launch was scrubbed that day and we returned as our booking demanded. WRESAT was launched the next day! It's just a pity that we didn't think to take the VW. (As you will read later, Richard did attend the launch—there are some inconsistencies in our recollections.)

WRESAT operated on its batteries for about five days before diving into the Atlantic Ocean on 10 January 1968 after 642 orbits. At least we can boast that Australis remains in orbit and will do so for 100,000 years more, give or take a few thousand!

Many years later, Dick Smith, the Australian explorer and electronics shop founder, discovered WRESAT's first stage in the South Australian desert. It was recovered and is now exhibited in the museum at Woomera. It was the last flight of a Redstone rocket.

* * *

Yet, 1967 was an even more significant year for me and, once again, the VW figured in events. Richard suggested a ski trip to Mount Buller and we would stay in the University Hut on the mountain. Now I was an experienced skier, so I jumped at the opportunity, which came with a condition – I was to drive the team to the ski field. And, it turned out, the team was to consist of Richard, me and those three girls from Parkville: Pauline, Delia and Ronnie. We were not going to fit five people in the VW without some considerable squeezing, but that was mere detail. As it turned out, Richard came down with a cold so the three girls and I set off for a weekend skiing, all squeezed into my VW beetle.

None of the girls were experienced skiers so it fell to me to help them. Pauline and Ronnie had spent an earlier weekend with Richard and Rex so at least they had a couple of days on the snow and knew what to expect. (I probably should not say that they were then notionally their respective girlfriends and boyfriends, so I won't.)

Delia had never even seen snow, so it was a totally new experience for her. Needless to say, the gallant knight attended her on the beginner's run, Bourke Street, until she fell awkwardly and decided to return to the hut, leaving me to try my hand on the Bull Run for experienced skiers. It was late in the season, snow sloppy, so it was not much of a day.

Plate 9.2

Plate 9.2 *My electronics designs were becoming more complex! This is the electronics package for an instrument to measure neutrons emanating from the sun.*

May I say that Richard's arrangements for us in the hut were . . . unusual, to say the least? Perhaps not for the 60s! We all slept in a bunk room and the wicked Delia insisted that the only male in the room kiss each of the girls good night. Perhaps I was mistaken, but it seemed to me that Delia's kiss was a little more passionate than the others.

Delia had fallen badly and her leg was hurting her so, given the less than perfect conditions, we began our return journey around midday, calling in at the Mansfield Hospital, where her leg was X-rayed. A small fracture.

I proposed to Delia in the early hours of New Year's Day, 1968, outside my family's home. She accepted and we were married on 11 April 1968. This was the best decision of my life. We are approaching our 50th wedding anniversary.

* * *

Diverted as I was by our blossoming romance, study called and weather satellite pictures had to be received while we waited for word from OSCAR on when Australis would be launched. And waited. And waited. It seemed that the US Air Force had lost their enthusiasm for flying some foreign satellite built by a bunch of school kids—no, University kids—and so Australis languished and languished in Lance Ginner's garage for two years.

I have to confess that as the 1968 University year got under way, Australis slipped further from my mind. Complex mathematics, fascinating electronics and preparations for our wedding came to occupy me.

It was my family's custom to go to a restaurant for lunch each Sunday and by that time Delia was regarded as a member of the family. On the Sunday after my twenty-second birthday, Delia and I arrived (in the VW, of course) at my family home to discover that my father had a very bad headache and we wouldn't be going to lunch. X-rays showed that it was an aneurysm[3] deep in his brain and nothing could be done for him. He died on 10 April, just three weeks before our planned wedding day.

Showing remarkable leadership, Delia arranged for our wedding to take place the next day with only our mothers and sisters attending. (Delia's father had died the year before under very similar circumstances, with the same consulting doctor and in the same hospital room.) We moved into my parent's home as we had planned, as my parents were due to travel to England for him to see his brothers and their families, but it did not happen. I feel I must record the actions of the Trinity College chaplain, Rev. Barry Marshall, who married us despite the traditional ban on all ceremonies after the altar had been stripped for Easter.

Life changed with early married life and as final exams approached. I passed with Honours and was left with a decision on what to do next. A year south with the Antarctic Division was out of the question now and Boeing Aircraft were no longer hiring Australian graduate engineers. By no means a last resort, I chose to join Professor Vic Hopper's department and study for a PhD in balloon-borne astronomy.

As previously mentioned, Vic was interested in balloons as a means of reaching above the Earth. He had formed a partnership with Professor Glenn Frye at the Case Western Reserve University (Case) to study gamma ray astronomy. As gamma rays are absorbed by a remarkably small amount of air, it is necessary to get above a large part of the atmosphere to measure gamma rays from a stellar origin. Balloons.

I was assigned (you were assigned in those days, at least in Vic Hopper's department) to study under Dr John Thomas, whose interest was infrared astronomy. Readers will not be surprised to learn that the wavelengths that John was interested in were absorbed by the atmosphere requiring telescopes to be carried above much of the atmosphere. Balloons.

---

3   A brain aneurysm is a ballooning of a blood vessel in the brain, and in my father's case it burst, eventually killing him.

Plate 9.3 (Opposite page) *Early morning launch of our gamma ray telescope from Palestine, Texas, December 1978.*

Large unmanned scientific balloons, of the kind we had tested Australis sub-systems on, were to be my life, and Delia's, too, as a balloon widow, for many years to come.

As I was an electronic engineer, no matter how inexperienced, I became the department's engineer and so was set design tasks in a very wide range of fields. I became a generalist with no specialist skills in any particular area of electronics. I will, however, describe two interesting electronic exploits of mine, among many.

Initial flights of the joint Case-Melbourne gamma ray instrument, a 2x1x1 metre, 200kg machine, showed that any gamma ray sources were

Plate 9.3

too weak to be detected by the relatively small (300x300mm) spark chamber detector. However, it was possible that a few pulsars[4] would be detectable, but this required that we measure the arrival time of every potential gamma ray event to an accuracy of one millisecond, and preferably much better. This on a balloon in a near-space environment with 10,000 volt sparks being set off routinely (this was part of the detector) and hundreds of kilometres from the launch site, often out of radio range. A system was devised but it did require synchronising with standard time (Greenwich Mean Time.) This was less than easy when our launch site was in Alice Springs in the centre of Australia.

There was one spark of hope—the super secret Pine Gap satellite receiving station was nearby. We managed to persuade a staff member to broadcast a timing signal, illegally, from Pine Gap. By the time the signal reached our receiver at the Alice Springs Airport, it was very weak, barely above the background noise. It took hours of careful scanning to find the very narrow pulse but once we had, we could synchronise the on-board clock and hence measure the time of arrival of our gamma rays to microsecond accuracy. This lead to extending the detection of pulsed radiation from the Crab and Vela pulsars by nearly a thousand-fold, from high X-ray energies (50 MeV) to medium gamma ray, perhaps 10 to 20 GeV.[5]

4   Pulsars are stars that emit electromagnetic radiation, over an extraordinarily wide range, in highly periodic pulses.

5   The energy of atomic particles are generally measured in electron volts (eV), so that MeV is the unit for a million electron volts and GeV for a billion electron volts. A 50 GeV gamma ray has the energy of an electron accelerated by an electric potential of 50 billion volts. One eV is approximately 1.6×10^{-19} joules, that is not a lot for you and me, but significant for an electron!

It is worth recording that among the more than a dozen balloon flights that I have been involved with, my group held two records—the highest flight of 156,000 feet (about 47km) and the longest of twelve days. No doubt both have long since been broken.

Later, while doing my post-doctoral year with Glenn Frye and his team at Case in Cleveland, Ohio, I was presented with the problem of packing many thousands of electronic detectors of what were essentially arrays of Geiger counters into a tiny space. The obvious answer was to have custom integrated electronics built, but this was well beyond our budget, so I devised a much simpler scheme based on available microelectronic chips. The idea, I was told, was later used on the NASA Compton Gamma Ray satellite, which carried a much bigger detector that greatly extended the number of known gamma ray sources.

Incidentally, Case was the site of the famous Michelson-Morley experiment that led to Einstein's theory of special relativity; that the speed of light is constant, irrespective of the speed that the observer is travelling or the direction she is looking.

* * *

Plate 9.4

So, life continued with Australis fading into the background. By mid-1969, I was in the midst of my PhD studies and a baby was on the way, due in January or February 1970.

I remember a singular day, one of those that one remembers forever. After the setback of a fire, the Apollo 1 capsule in 1967 in which three astronauts were killed, Apollo 11 was to be the first manned landing on the moon. It seemed that everyone in the world was glued to television sets to watch the launch on 16 July, land on 20 July, and then, at home for lunch, Delia and I watched Neil Armstrong step down off the spidery

Plate 9.4 *The improved gamma ray detector, a multi-wire proportional counter. There are over 500 of my "electronic detectors" visible in the image (but very hard to see) on each of the eight trays of proportional counters. Each tray was about one metre square.*

lunar lander and onto the surface. Buzz Aldrin followed him twenty minutes later. Despite the poor black-and-white quality video, which improved markedly with later flights, the moment was truly memorable and historic. Each successive landing produced better video quality and the final flight, Apollo 17, even showed the lift-off of the Lunar Module's return to lunar orbit stage *in colour!*

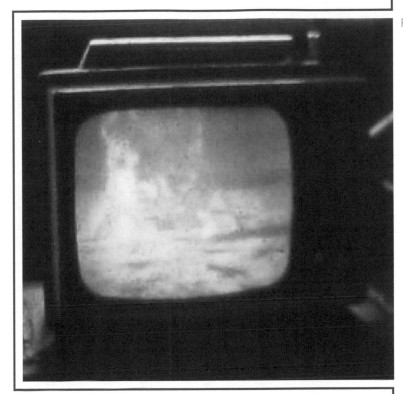

Plate 9.5

It is interesting to learn that the origins of the Apollo program began during President Eisenhower's administration[6]. It was Soviet Russia's space feats, including the manned space flight of Yuri Gargarin on 12 April 1961,[7] and political pressure to ensure American superiority over the Soviet Union that spurred the new President, John Fitzgerald Kennedy, finally to commit the US to its manned lunar program.

Memories of Australis receded until . . .

---

6   Wikipedia, 'Apollo program' at <https://en.vikipedia.org/wiki/Apollo_program>. Accessed 31 July 2017.
7   Wikipedia, 'Yuri Gargarin' at <https://en.vikipedia.org/wiki/Yuri_Gargarin>. Accessed 31 July 2017.

Plate 9.5 *Apollo 11 moon landing as seen on our small black and white television at home. One astronaut is clearly visible in the foreground of this fuzzy image, another on the left in front of the lunar lander.*

Plate 10.1 *Australis OSCAR A underging testing at NASA Goddard. From the left the HF transmitter, space for one of the ejection springs, the HI keyer and the VHF transmitter are visible with the battery packs behind.*

CHAPTER 10
The Flight

Recall that Jan King was on the board of directors of AMSAT Incorporated, the east coast group who were interested in continuing the work of Project OSCAR once it became clear that US Air Force launches from the Vandenberg Air Force Base were most unlikely to be repeated. This group had senior technical and managerial people, as well as the junior engineer, Jan King, who was given the task of preparing Australis for launch. All were radio amateurs and all had day jobs with NASA, companies and universities.

While there was a German project progressing, and it became OSCAR 7, they realised that Australis might be their entry into amateur radio press. As Jan recalls, 'We knew only the general story because it had been reported (in the general and amateur radio press). Remember that Owen and Richard had done a very good job of publicising what they'd done. They were on radio and it had been in any number of articles, including the major amateur radio magazine called *QST*.' Thanks, Jan.

It made sense for them to take over Australis from Project OSCAR. And so, it was shipped to Washington in late 1968 or early 1969.

Jan continued, 'Our impression of the spacecraft itself was – I was looking at a lot of flight hardware every day – considered to be all right, although the (outside) case didn't quite look up to the job. However, when we managed to turn it on and operate it we realised that a lot of work had actually gone into it … We did refurbish a number of things (including the batteries) but it's fair to say that the primary sub-systems they'd developed were flown pretty much as is. We had to do a couple of things we found in testing where something wasn't quite right and had to change a couple of components for better quality as they didn't have access to the kind of parts that were sitting in a drawer downstairs and I could just walk down, sign a little piece of paper and take the hardware. That was all I had to do to get the parts.' I understand that additional filtering was required on the transmitters.

It is well worth recalling what was happening in 1969 when Jan and AMSAT were testing Australis in preparation for a launch. Apollo 11 had landed on the moon in July 1969 with the ever-memorable remark, 'That's one small step for a man, one giant leap for mankind', and Apollo 12 in November 1969. It was not yet time for the disastrous explosion of Apollo 13 in April 1970. NASA was completely taken up with the Apollo program and then there was this annoyance of a foreign satellite wanting a piggyback ride on a NASA flight.

Following filings with the Federal Communications Commission (FCC), there were, as Jan describes it, two political problems and a technical one. Jan was more than capable of handling the technical issues we've already described—he was a radio amateur, after all.

POLITICAL ISSUE NUMBER ONE

Australis was, of course, not an American satellite and there was no precedent for flying foreign secondary payloads on US government missions. Furthermore, this Australis satellite had been built by non-professionals with no experience in building spacecraft. However, Americans were working with the Australians, so there's professional American involvement.

By that stage, NASA had flown secondary payloads to test the Apollo network, so there was precedent for flying secondary payloads and, besides, the US Air Force had flown the four OSCAR satellites as secondary payloads. So, if the Air Force can do it, so can NASA. Besides, it was an international program with a friendly and supportive country that trained young people 'so there's education involved in the process'. They had a good story for NASA decision makers. Political problem number one pretty much solved, so now for . . .

POLITICAL ISSUE NUMBER TWO

Who owned the flight and, thus, who approves flying the secondary payload?

Now, this is not as straightforward as it might seem. The proposed flight was with TIROS N, which was a program of the National Oceanographic and Atmospheric Administration, NOAA, but the launch rocket was owned by NASA. As Jan succinctly put it, 'They (NOAA) don't own the rocket, they own the ride' and the spacecraft. NASA owned the rocket. So, who approves a secondary payload? This was a new problem that had not been encountered before—recall that the US Air Force owned both the satellites and the launch vehicles, so they had complete control over their flights.

At that moment, NOAA was transitioning their weather satellites from the development of a system to help weather forecasters to an operational system. TIROS M was to be a proof of concept for the operational system and so Australis was to fly with the first operational satellite and an improved weather satellite system.

To further complicate things, the NOAA satellite was to be launched on a Delta rocket. The Delta was to become the standard launch vehicle for government satellites. The Delta project was continually improving performance and the vehicle for the TIROS launch had greater performance, and greater complexity and risk, than was necessary for the TIROS flight. Then add the risk of a foreign secondary satellite. Using AMSAT's board's connections, a presentation was made to the satellite managers and at the end of the hour-and-a-half, the response was, 'I have three responses to your presentation in terms of how ITOS feels about this mission . . . "no, hell no and never". You can take your pick from those.' Most people would interpret that as negative!

However, Jan was encouraged by the smile of a Delta rocket person (John Tomasello) in the room. Once again, the amateur radio mafia was called upon to introduce him to the NASA Deputy Director for science, Dr Dick Marsden. Yes, Marsden was an amateur. Marsden decreed that Australis was a science mission and, therefore, he had jurisdiction. Furthermore, he said yes, NASA would fly it on the Delta. There was one provision, however. NASA would pick up the tab! No cost to AMSAT!

An AMSAT member, Dick Daniels, worked in the NASA headquarters and so he was able to place Marsden's letter of approval before Thomas Paine, NASA's Administrator, who signed it. By the time Payne had signed the letter approving the secondary payload, the Delta project was already preparing hardware to mount Australis on the second stage.

One of Paine's stipulations was that they had to get permission from the Deputy Associate Administrator of NOAA, Jack Townsend. That presented few problems since Townsend was … yes, a radio amateur. However, Townsend required an entire environmental test program, which was undertaken—mostly.

Jan describes the laboratory behind another in the basement of the Goddard building where Australis was refurbished and prepared for flight by a group of 'bright young engineers'. Their managers gave them the freedom to do what they wanted to do, realising that 'it wasn't exactly in line with our position description at work. This was teaching us something that we couldn't get from everyday engineering', recalls Jan.

Among the 'everyday engineering' that Jan was employed to do was testing one of the lunar rovers used by the astronauts and seen by countless millions around the world as they drove (or perhaps hooned?) around the moon's surface. At the time, there were fifteen spacecraft in the building being tested, all multimillion-dollar missions, and here they were playing with this tiny Australis project.

One of the obvious differences between Jan's experience at NASA Goddard and ours in Melbourne was the enormity of the difference in resources of every kind: knowledge, experience, intellectual and physical. The only resource we had other than our own knowledge gained through study and amateur radio was the University's Bailleau library. As the reader can imagine, less than a decade after the launch of the first satellite, there was precious little in the way of technical literature in the library. Especially so, since at the time the Soviet Union and the US were in the midst of the Cold War and almost any information about rockets and satellites was considered to be of military significance.

An obvious modification that was performed was to apply a thermal layer to the outside skin. When built, we knew that some sort of surface treatment would be necessary to control Australis' internal temperature but had no idea whatsoever of what this might look like. We painted the Australis model we used for publicity in black-and-white stripes, not to recognise any allegiances to any football team but to mimic OSCARs 1 and 2. The surface applied at Goddard was a matt black over much of the surface area and with a square of gold foil around the antennas, giving the appearance familiar to us now. Just before the launch the rubber compound was removed to reveal a bright chrome surface.

Jan mentions that he is unsure whether the springs were replaced. I can't think why they would need to be, so I will say, tongue in cheek, that because Australis OSCAR 5 (as it is known once in orbit) entered orbit with a relatively low spin, they must have been the Australian springs!

* * *

It's worth mentioning that I have a distinct memory of being told, perhaps by Willson Hunter, that approval for the Australis launch had to be obtained from the political head of NASA, the Vice President who in 1969 was Spiro Agnew. Nobody else confirms that memory.

* * *

Once we had word of the possibility of a launch and later as it became more certain, MUAS began to mobilise. I dusted off my antenna pointing prediction computer program, which I called Standard Orbits, and began to write a Users' Guide.

In 1969, there were no copying machines as we know them today. No word processors, only typists, mainly women, so my first task was to persuade Maeve, Vic Hopper's secretary and the department's typist. Yes, she would type the Users' Guide for me without charge but we had to be careful because, recall, Vic Hopper wasn't favourably disposed to this "silly satellite stuff". Next, we wanted some pictures and calibration graphs, so a small number of lithographic plates were donated.

The frontispiece carried pictures of the striped Australis and a view of its internals. The first page of the Guide, which is reproduced in Appendix A, carried the words:

> The Australis OSCAR A Users' Guide contains full instructions for all wishing to track the satellite. Since the success of the project depends on the support of a large number of tracking stations, we are anxious to enlist the co-operation of suitably equipped radio operators, short-wave listeners and VHF enthusiasts everywhere.

The opening paragraph of the first chapter, Background, is interesting to view in the light of today's knowledge:

> The Melbourne University Astronautical Society was formed at a time when the image of space research was dominated by a spirit of adventure. Today, much of the popular interest has subsided, but the potential of spacecraft is being rapidly revealed. The satellite is an indispensable tool in many fields of research; its use in communications, navigation and meteorology is commonplace.

This, with the Apollo moon landings not far away!

The second chapter described each of the sub-systems in a similar manner, but less detailed, with block diagrams of the electronics and a drawing showing the configuration of each of the sub-systems within the satellite.

Chapter Three contained a description of organisation of Regional Directors for each of the three International Telecommunications Union regions and Local Co-ordinators for Asia and Australia, including the Australian states, Japan, Malaysia and New Zealand. Next followed instructions for the co-ordinators to provide antenna pointing directions to users within their region. These included a sample print-out from the IBM 7044 computer – I was concerned that I might be reprimanded for printing off hundreds of pages of the same figures that were part of the Users' Guide. I was not reprimanded, perhaps because of certain blind eyes, or perhaps the print-out was not noticed.

The final chapter showed how to receive the telemetry signal and how to use the calibration graphs provided in the Guide to determine battery voltage and current, temperatures and the satellite spin. A sample completed coding form was included.

Finally, there was a page of acknowledgements of donors.

Several hundred Users' Guides were roneo'd[1] and distributed to amateurs via the co-ordinators. It is reproduced in Appendix A.

*　*　*

The press was alerted and articles published in magazines to ensure that as many amateurs as possible were excited and prepared for the flight.

*　*　*

Some weeks before the launch, I was telling my wife, Delia, about the satellite and, seeing my excitement, she volunteered that she would like to see where all this happened, that is, our meeting room. It should be mentioned that she was very near to term with our son, Nicholas. Off we went one lunchtime to the University. (Thankfully by this time I was a very junior staff member and had a car park nearby.) Into the Natural Philosophy Building, up the stairs, out onto the roof to be presented with the near-vertical ladder to our meeting room. Heavily pregnant, Delia was not put off by this obstacle and she bravely struggled up the ladder and into the tiny room, much to the surprise of those inside. Nevertheless, she saw something of what the excitement was about before climbing down the ladder and returning home.

Our son, Nicholas, was launched two-and-a-half weeks after Australis. Not good timing for a newly arrived baby and a satellite!

*　*　*

As the launch approached, a regular 'sked' on amateur radio was established, we at Les Jenkins' home and Jan and his team in Washington. The launch date narrowed as we got closer to the launch and the initial orbit parameters for TIROS were made available. Off I went and eagerly ran my Fortran program again to produce the 'Standard Orbits' for TIROS and, hence, for our satellite. Once the launch date became fixed, the information was disseminated.

1　　A relatively cheap, but antiquated and dirty, means of producing multiple copies of a document from a master sheet that had been cut by a typewriter. The typist had to hit the keys heavily to ensure that the master was cut through so that the printing ink could flow through it to the paper. Considering the technology, it produced a remarkably good quality product.

Let us step back for a moment from the preparations that we were making in Melbourne. As Jan acknowledged, our publicity had been a major part of the Australis project and we were not going to let it slip at this point. We wanted to invite the press, contributors and friends to the launch, as impractical as it may sound. However, if we could re-broadcast a commentary from the control room of the Delta-TIROS launch into a room in the University, that would have to suffice. In those far-off days, interstate telephone calls were expensive and international calls were simply unaffordable. So, once again off to the telephone system provider, in those days the Post Master General's Department (PMG), a federal government monopoly of postal and telephone services.

I have no recollection of how we managed to contact the appropriate people in the PMG, although I had done a Christmas holiday work placement in the Department so perhaps I used my contact there. Steve was, I think, working for the PMG by then, so he may have made the contact. However it was done, we managed to persuade the PMG to give us a line between Washington and Melbourne for one hour on the evening of the launch. The normal charge would have been $600 for that hour!

Back to Washington for the launch. Jan and his AMSAT colleagues were invited to the Mission Director's Centre in Washington for the launch. He recalls the moment that our satellite separated from the Thor Delta rocket and it became Australis OSCAR 5. They were intently watching a strip chart recorder[2] when the pen moved producing a blip on the paper – this was the moment of separation. The chief technical officer at NASA congratulated Jan. What a moment that must have been for a relatively junior engineer who had been moving in the highest circles of the organisation.

Back to Melbourne where the telephone line in our meeting

Plate 10.2

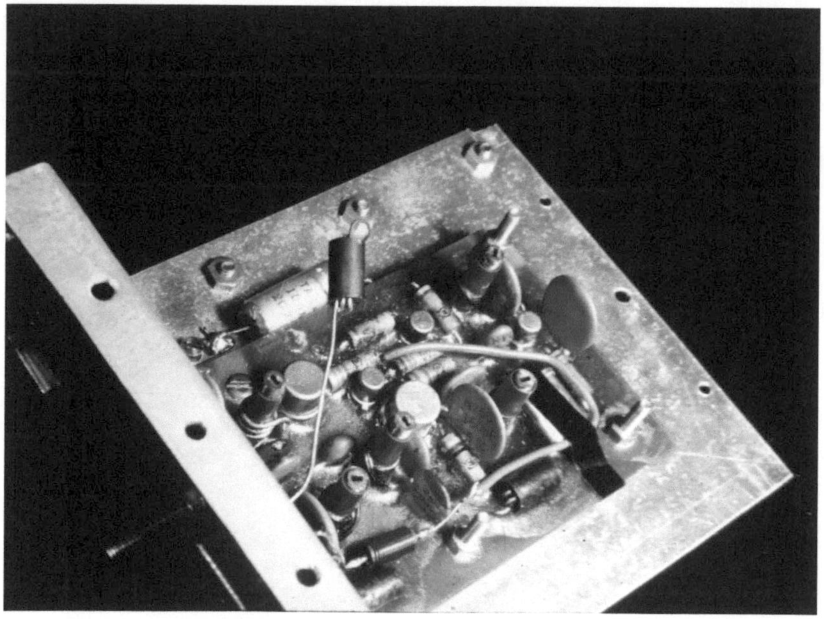

---

2   A strip chart recorder is a machine with a steadily moving roll of paper with a motor driven pen that writes on the paper strip, thereby recording changes to an electrical signal, for example.

Plate 10.2 *The command receiver being modified.*

room garret was extended into the lecture theatre below and an amplifier rigged up to broadcast the commentary into the room. (That an ordinary person dared to meddle with the PMG's telephone lines and actually broadcast the conversation was undoubtedly illegal. Never mind.) I had prepared a large world map and sketched the first few orbits onto it, highlighting it so that everyone in the theatre could see the track. I am looking at that map as I write this. The telephone connection was made perhaps half an hour before the launch as the press, friends and members gathered for the launch itself.

The drama and tension escalated as the countdown proceeded.

'First Stage?'

'Go.'

'Second Stage?'

'Go.'

'Satellite?'

'Go.'

Then the countdown with everyone in the room holding their breath. Fingers crossed, hoping.

'. . . three, two, one' an eternity waiting, then 'lift off'. Our baby was on her way into space! It was 9:37pm on Friday 23 January 1970, in Melbourne, a moment few of us in MUAS will forget.

Jan recalls the cheers that erupted when the TIROS satellite separated from the second stage Delta rocket. We heard those cheers in Melbourne and the later cheers as Australis was gently pushed into its own orbit by the Henderson springs, the micro-switches operating and connecting the batteries to the electronics. She was now to be known as Australis OSCAR 5, or AO5.

The launch took place from the Vandenberg Air Force Base in California that we had visited two-and-a-half years previously and the launch was near perfect, as close to the design orbit as it could be. We traced AO5's path from Vandenberg on the west coast of the US southwards over the Pacific Ocean, over the top of Mururoa Atoll where the French had conducted their nuclear weapons tests, toward Antarctica then northwards between Madagascar and the African continent. Crossing the equator, she flew along Italy, over Norway and Greenland, before turning south again and passing over Canada, the Cook Islands and onwards. It was not for some time that we learned that AO5 had been heard soon after separation by a radio amateur in Madagascar.

One-and-a-half hours after the beginning of the telephone call, we hung up, although the PMG people did not seem to be in a hurry to finish the call, even if the cost would now have been around $900!

At the same time, Les had been monitoring AMSAT broadcasts from his radio shack in Cheltenham. It was not until the second orbit that a New Zealand amateur heard AO5 and some three-and-a-half hours after launch that the signal was picked up in Australia. It was a weak signal at a low elevation as AO5 passed southward down the Tasman Sea between New Zealand and Australia. The next orbit passed southward over Australia, over Townsville, Queensland, and Mount Gambia, South Australia, some five hours after launch. A few dedicated amateurs heard AO5 at around 2:30am on Saturday morning.

* * *

Here is Rod's recollection of the launch:

'I was a PhD student when the satellite was launched, and that year I was President of the Astronautical Society. We invited a contingent of reporters from the local press to a lecture theatre in the Physics Building, close to the Astronautical Society clubrooms. A telephone hook-up to the launch site was organised and the countdown was relayed over the phone, as Owen described.

'At our end of the phone hook-up, Richard Tonkin communicated with the launch site and provided his own commentary. We tapped into the phone line and played the commentary from the US, together with Richard's commentary over the public address system in the lecture theatre. Unfortunately, Richard's voice was much louder over the public address system than the weak voice from the US. We had to ask Richard to talk quietly, so that we could adjust the volume to a level that enabled the voice from the US to be heard without Richard's voice dominating. But as the launch proceeded, Richard became more and more excited and voluble. He more or less completely drowned out the commentary from the US. It was fantastic to watch Richard's excitement (which we all shared). There was no way to quieten him down!'

* * *

And here is Richard's recollection of the launch:

'Owen has described Australis in detail and with much greater clarity than I could recall. Suffice to say that, some two years after Peter Hammer and I laboriously tapped out our letter to Project OSCAR on someone's manual typewriter, Australis was a reality and we took it to California in June 1967, hopeful of a launch into orbit within a few months.

'In fact, it took two further years before Thor/Delta Number 76 soared off the launch pad at Vandenberg Air Force Base, California, on 23 January 1970, into a foggy morning sky. The Delta's main satellite was TIROS-M, or ITOS-1 when it reached orbit, an advanced polar orbiting weather satellite that NASA was launching for the National Oceanic and Atmospheric Administration (NOAA)—effectively the US Weather Bureau. I am sure that the scientists and engineers at NASA and NOAA anxiously monitored the progress of the launch, but not as anxiously as the little group that had built Australis, who were huddled in what the local media called 'their tiny rooftop garret', the little room in the attic of the Physics Building that housed, apart from our tracking and recording equipment, a big, noisy electric contraption that pumped air (not air conditioning) into the windowless main Physics Lecture Theatre below.

'We had reason to be anxious—the Delta rocket had a patchy success record; of the four previous launches, two had failed, and there was no Australis 2. We had no back-up in case we ended at the bottom of the Eastern Pacific Ocean. This was our one and only shot at getting our satellite into space and realising what we had lived and dreamed about for five years.

'The predecessor to Telstra, the PMG, had kindly provided a telephone connection between our 'tiny rooftop garret' and the AMSAT-OSCAR people at Vandenberg, so we had a blow-by-blow description of the launch. The solid rocket boosters that circled the first stage of the Delta burnt out on schedule and dropped away. The first stage Thor ballistic missile shut down as planned, the second stage separated and ignited. So far so good, but that was what the man who fell out of the 15th story window said as he passed the second story on the way down—plenty of things could still go wrong. But Delta 76 was on its way and it kept going up. The fairing, which protected the satellites until they were out of the atmosphere, came off as planned. The second stage continued to burn

steadily. 'ITOS separation!' we heard over the phone line. 'The orbit looks good.' So the weather satellite had been safely delivered into space. Now we had to wait. ITOS needed to be far enough away from the burnt-out second stage so that there was no danger of it colliding with NOAA's brand new satellite, and Australis could not be pushed away from the rocket with its two Henderson Federal Spring Works separation springs until that distance had been achieved. We waited— anxiously is too mild a word—for what seemed like hours, but was actually only a few minutes, for news from the AMSAT-OSCAR team at the launch site. It was so quiet in our control room atop

Plate 10.3

Plate 10.3 *The command receiver being modified.*

the Physics Building (we had made sure that the 'air conditioner' electric motor was switched off) that it seemed we had all stopped breathing. . . perhaps we had! Then, the beautiful words we had waited so long to hear, 'NASA has just received telemetry from the Delta second stage—Australis has separated and is in a stable, circular 1,400 kilometre polar orbit'.

'Five years of pent-up emotions erupted in that 'tiny rooftop garret'—there were shouts of joy, back slapping, hugs and a few tears. We had done it—a bunch of scruffy undergraduate students had designed and built a satellite from scratch. It was safely in orbit and it worked.

'I could not have imagined, as I watched that faint star, Sputnik, glide across the night sky in October 1957, that a little over 12 years later I would look up at that same sky and know that I was a proud member of the team that had made our own satellite.'

* * *

I wrote an Interim Report of the flight of AO5 and here it is reproduced in full, with my typing and spelling errors included, and in typewriter typeface:

```
AUSTRALIS OSCAR 5 INTERIM REPORT. by Owen Mace.

Australis OSCAR 5 is now silent, its batteries discharged after a working
life of six weeks. The work of collecting and processing the thousands of
reports from amateurs around the world begin. Before the last report on AO5
is written, many hours of computer time will have been used in the processing
the data received from amateurs. Concurrently with this work, the next
Australis satellite is being planned and designed.

AO5 was launched at 1131 GMT on January 23rd. in what could only be described
as a fl awless, text book launch. One hour later, Australis separated from
the Delta second stage and its two transmitters switched on. 5R8AS reported
hearing the VHF beacon a few minutes later as the satellite came into range
of his Malagasy Republic QTH. Minutes later as AO5 passed over Europe, DJ4ZCA
and DL3OJ heard the 10 m beacon. In the following few orbits Australis was
heard by amateurs in the UK, US, New Zealand, and, of course, Australia.

The response of US amateurs, especially, is staggering. Many thousands of
reports have been received by the project to date. Some tracked every orbit
in range throughout the life of AO5, some reported extraordinary antipodal
propagation effects, and one even correlated the horizon sensor signals with
cloud formations derived from weather satellite pictures, during the later
part of the VHF transmitter's life. The patience of one brave soul is attested
by this log entry:
```

"On orbit 181/182, I could hear the 10 metre signal just about all the way around the world. I heard it for 95 of the 115 minute orbit, from very faint to fairly strong signal strengths."

WA2KS heard the 29 MHz transmitter commanded off during orbit 61 on January 28thÂ.

At the Project Australis headquarters station, VK3AVF[3], teams were organised to track the two high elevation passes each morning and afternoon. This vigil was maintained until the VHF beacon ceased transmissions during orbit 230 on Saturday 14th. February after 3 and a half weeks of highly successful operation.

The magnetic attitude stabilization system (MASS) worked very well also. The satellite was soon locked to the earth's magnetic field by the MASS magnet. So accurate was this tracking that, by the 10th. February, the signal strength of the VHF beacon was appreciably lower as the satellite was south of the city, than when it was north. This was caused by the transmitting antenna becoming unfavourably directed by the earth's field as the satellite moved. Future designs will undoubtedly allow for this unplanned tracking accuracy!

The accompanying article by Jan King describes some of the preliminary results from Australis. In the ensuing months Project Australis will be analysing all the reports received in considerable detail to determine the affectiveness of the design procedures in order to incorporate modifications to the next satellite. Any reports are welcomed, so, if you have not already sent in your reception reports and station resume, please do so. The address is

> Project Australis(Telemetry)
> c/o Melbourne University Astronautical Society,
> Union House,
> University of Melbourne,
> Parkville, Victoria, 3052

* * *

---

3   Radio amateurs are assigned call signs with the first letters signifying the country of their origin: VK for Australia; W for the USA, etc.

The report went on to discuss plans for our next satellite. The remainder of my report is included in the next chapter.

Internal temperatures began at about 24C and rose steadily until it had stabilised between 47C and 49C by about orbit 78. These temperatures were somewhat higher than we had hoped for but well within the range that the electronics were able to cope with. The skin temperature, on the other hand, varied considerably as AO5 rotated in orbit from 45C to as much as 63C.

The spin rate, as estimated by the horizon sensors, was around 3.5 revolutions per minute during the first few orbits. After two weeks in orbit, the spin rate was so slow that it did not complete one revolution in a pass of twenty or so minutes! By this stage, AO5 was travelling with the VHF and command antennas roughly aligned north-south and in the direction of travel, as planned, so that fading of the VHF signal was reduced. The spin caused the HF antenna to rotate slowly and so there were pronounced 'nulls' in the HF signal strength.

We were surprised at comments reporting that the horizon sensors recorded clouds, snowy mountains and registered transitions of the sun and even the moon!

Plate 10.4 *A US Government Interagency Motor Pool car 'For Official Use Only' delivering Australis OSCAR A to NASA Goddard for testing.*
Plate 10.5 *Australis OSCAR A about to be tested on a vibration table to ensure that it could survive the launch. Photograph courtesy of NASA.*

What I failed to mention was that there were failures, or at least some difficulties. First, the telemetry modulation on the 10-metre HF beacon 'dropped off sharply' at the end of the first orbit, making it very difficult to decode the data from the HF transmissions. (It was also reported that the failure occurred on the third orbit. No matter, modulation was much reduced for much of the flight.) It did appear about a month into the flight. The cause is unknown but my guess is a soldered joint somewhere had been shaken loose and a temperature change reconnected it later.

The Command System required considerably more radio power to activate it than had been calculated. Nevertheless, the HF transmitter was successfully commanded several times amidst much trepidation that we might not be able to turn it on again.

I was a little disappointed with AO5's operational life as I had hoped for a two- to three-month useful life, rather than the approximately 580 orbits over six weeks. This was largely due to the difficulties with the Command System and thus the 'high power' HF transmitter being left on for longer periods than we had planned.

Appendix B reproduces Jan King's Interim Report, which contains important information about the flight, as well as references to other articles and papers.

Plate 10.6 *Australis OSCAR A in its fighting uniform, the thermal control surface treatment. Note the locating pins, spring wells and tape antennas.*

Plate 10.7 *Australis OSCAR A mounted on its 'baseplate' with antennas wrapped around it ready for mounting in the launch rocket.*

* * *

We might mention that we have pictures showing Australis was mounted very close to the second stage rocket motor. Perhaps this was to ensure that the NOAA people were satisfied that this non-professional, foreign satellite was well away from their precious satellite. In any event, this would not have been a comfortable place to be, less than a metre from the shaking and thundering of the rocket engine.

Plate 10.8

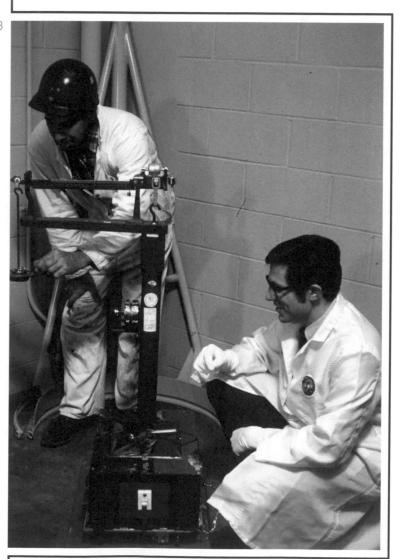

Details of the Thor/Delta rocket, TIROS and the orbit are given in Appendix B.

* * *

It is worth imagining how things would have turned out if it had been Apollo 12 that had the in-flight explosion in November 1969 before the launch of Australis and not Apollo 13 after the Australis flight. The senior men in charge of NASA were the very same ones that approved the Australis launch and soon after dealt with the Apollo disaster.

Plate 10.8 *The pre-flight, weigh in. Jean Shrimpton, please note the white gloves!*
OPPOSITE PAGE
Plate 10.9 *Launch! The Thor/Delta takes off into the mist from Vandenberg Air Force Base on 23 January 1970 with Australis OSCAR 5 on board. The University of Melbourne Gazette, Vol. 26, Number 2, 15 April, 1970. Reprinted with permission.*

UNIVERSITY OF MELBOURNE
GAZETTE

Published for University circulation, as occasion requires, by the Registrar for the Council of the University of Melbourne

Vol. XXVI, No. 2 — Melbourne, 15 April 1970 — *Registered at the General Post Office, Melbourne, for transmission by post as a periodical.*

Student Satellite in Orbit

10.9

Thor-delta 76 blasts off — Photograph, *NASA*

On 23 January 1970 the seventy-sixth Thor-Delta rocket was fired from the Vandenburg launch pad in California. The primary spacecraft was a NASA weather satellite, TIROS-M. However, to a group of students at Melbourne University the secondary satellite was of much more interest. This was AUSTRALIS OSCAR-V, a satellite designed and built by members of the University Astronautical Society. As AUSTRALIS OSCAR-V separated from the Delta second stage, two radio transmitters were turned on, and their antennae unfolded. About one thousand miles beneath the satellite, an amateur radio station in Malagasy, reported that the very high frequency (VHF) transmitter was operating. A few minutes later a weak signal from the high

10.10

10.11

Plate 10.10 *Ready to launch. Front page of the Goddard News magazine dated 9 February 1970. Credit: NASA.*

Plate 10.11 *Celebrations a few hours after the launch. Note the time on the wall clock.*

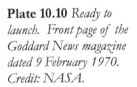

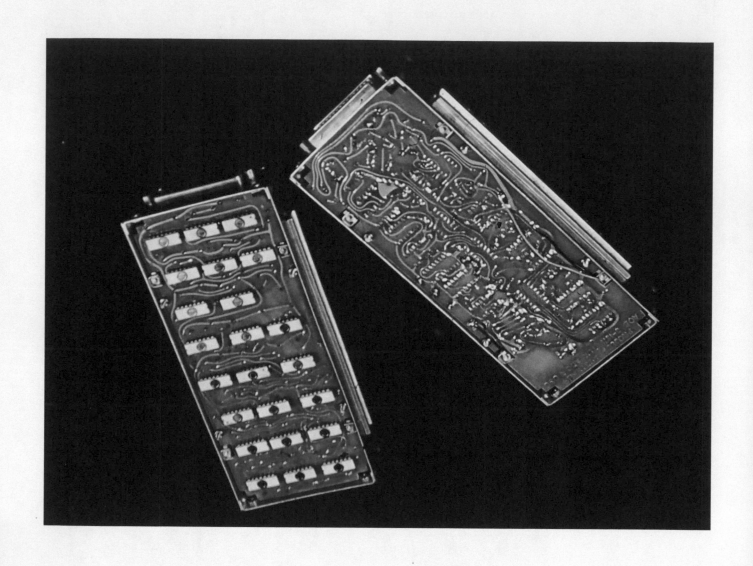

Plate 11.1 *Technology had moved on for Peter's radio teletype (RTTY) telemetry circuit boards for OSCAR 7 – CMOS logic chips – still working 40 years after launch.*

CHAPTER 11
Our Lives in Space

As noted in the previous chapter, I wrote an Interim Report on AO5's flight. The second half of the report described what we were thinking following the last transmission of our satellite. There is no date on the report, unfortunately. Once again, a typewriter typeface is used and all spelling and grammar errors are included to evoke the memories of those days:

Work is proceding with the design and testing of the next Australis. It is envisaged that there will be seven main areas of work; subsystems;

The main experiment, the repeater. While other groups, notably DJ4ZCA, are strongly in favour of linear translators, it is felt by the Australis group that the next step ought to be a hard-limiting FM system. Thus Project Australis is working towards a multichannel, channelised FM repeater. The plan is to use one receiver to mix to some convenient IF stage, split to several separated IF filters and detect down to baseband. Then each demodulated signals can then be used to frequency modulate its own carrier which is amplified and radiated. This system allows a number of advantages in that signal processing of the baseband is possible (eg speech compressing)and ALC) as well as removing doppler shift from the upgoing signal. It is presently planned to use about six channels receiving on 2m and transmitting on 432 MHz.

Telemetry System. It is hoped that a 60 channel telemetry system will be accomodated on the next satellite. Its output will be in the form of teletype signals impressed on one of the repeater transmit channels and operated on command.

Command System. A 35 channel command system will be incorporated to allow switching of receivers and transmit channels. This will allow great flexibility and will allow failed subsystems to be removed from the repeater system.

MASS. It is possible that a magnetic system similar to that carried by AO5 will be incorporated, although a gravity gradient stabilization has been mooted by Amsat.

Power Supply. A 6 watt solar powered battery is under investigation by Amsat, who will be responsible for the power supply.

Package. This is the responsibility of Amsat.

The design is for a lifetime of one year, to 85% confidence. It is anticipated that a prototype of the system will be flown on a balloon from Mildura in the near future, and interested amateurs are asked to listen to their broadcasts for further details,

Clearly, we had a much more sophisticated satellite in our minds and Peter takes up the story of Australian satellite building below. However, my family and my PhD studies drew me away from MUAS. I was sent on my first overseas field trip in 1971 to the National Science Balloon Facility in Texas. Imagine a three-month overseas trip these days without family, including a one-year-old infant! This was not the first long field trip, nor the last overseas field trip.

However, I was lucky enough to be seconded with Delia and Nicholas to Prof Glenn Frye of the Physics Department at Case Western Reserve University for my postdoctoral year in 1973. What a year it was for us all. At Case, we were building a totally new gamma ray telescope[1], which was much, much bigger than the previous one. Glenn and Ruth Frye, their friends and colleagues, were extraordinarily hospitable and it was a fabulous year for us all.

A memorable moment came for me when we were discussing a power supply for the new instrument. Batteries were considered but, being fascinated by the space shuttle that was being contemplated then, I knew that fuel cells were being studied. Glenn told me to make a cross-country telephone call to California where the design studies were being conducted. I spoke to NASA's top expert on fuel cells, who spent an hour with me discussing our requirement and the technology! Together, we decided that batteries were our best solution and I discovered later that the Shuttle fuel cells cost many, many millions and so were utterly out of the question for us.

There have been a surprising number of space-related projects that I have had some connection with. Rather than describing each in detail, I will select just two.

The mid- to late-1980s saw an Australian Space Office formed to bring Australia into the space age. Among its many activities was to arrange for a group of Australians with experience and influence in space matters to visit the Russian Space Institute in Moscow and see some of their projects. The entire team was warned by 'defence security' to be on our guard against being blackmailed for an indiscretion. The irony was that President Regan was visiting Premier Gorbachev at the same time as our visit and, by and large, the Russian security services had many other much more important matters to concern themselves with.

We were all warned that, should we return to our room after dinner to find an attractive naked woman in our bed, leave immediately and report the matter to the Russian-speaking keeper of our room keys. As we got to know each other during the visit, we began to ask each other over our caviar breakfast

---

1   Called a Multi-Wire Proportional Counter (MWPC) with the detector being ten times the area of the older 'spark chamber'. I understand that one of Compton's gamma ray detectors used a similar detector of similar size.

(literally) if anyone had met 'Natasha' the previous night. There was general agreement, if not disappointment, that the answer was no.

The warnings were not without reason, however. One evening, I was sitting in a lounge and pulled up my wing back chair to hear the conversation more clearly. It took a bit of force to move the chair and to break the microphone wire!

Nevertheless, it was a most interesting and indeed enjoyable ten days in Moscow.

The Space Office continued for over a decade and among the contracts that it funded was a project to build test equipment for the Advanced Along Track Scanning Radiometer (AATSR) that flew on the European Envisat satellite (2002 to 2012). This instrument, which had been co-invented by another of Vic Hopper's PhD students, was the latest in a series of similar instruments to measure sea surface temperatures from space. The significance of the measurement is that sea surface temperatures can be measured from space to an accuracy of around 0.3C over a much wider area than is possible by any other method and so represents a figure for the average temperature of the Earth itself. Furthermore, the multi-decade series of measurements are highly significant for climate research.

Much to the chagrin of the 'traditional' winners of such space projects, the company that I was working for, Aspect Computing, won the contract for the AATSR Ground Test equipment. To further add to the annoyance of our competitors, we did so profitably and on time. The state manager of Aspect encouraged numbers of successful bids for projects that would not ordinarily have been the province of an information technology company and yet were largely software production, which was our strength.

I have been extraordinarily lucky in my professional life to have seen and done all that I have. (Should I mention a trans-Atlantic Concorde flight as part of a business trip?)

Let us now read how others in our group have lived their lives.

* * *

RICHARD TONKIN

At the time of the launch, I had been married to my wife Pauline for just over a year. We were living in a flat in East Malvern, she was pursuing her musical career with the Melbourne Symphony Orchestra and I was working full-time at the Weather Bureau in Melbourne, in charge of receiving and processing pictures from American weather satellites. Frankly, I was over Australis, Pauline was pregnant with our first child, Karen, by October that year and we had bought a block of land in Ringwood and started to build a house.

So, I did not follow Peter's exploits with OSCARs 6 and 7—perhaps if some of the Australis team had stayed together, we could have gone on to create a business building small satellites, negotiating launches and selling space on our satellites to the universities and the CSIRO for experiments. But it was not to be.

In 1972, I finally finished my Law course, qualified as a lawyer and got a job with a firm in the northern suburbs of Melbourne, where I recently retired as the senior partner.

Satellites were still in my DNA. We moved from Ringwood to acreage in Smiths Gully, north of Melbourne, had two more children, I worked hard in the law firm, as did Pauline in her musical career. My interest in space was renewed when I joined a tour led by an American group to Russia in 1992 to watch a manned Soyuz launch to the Russian space station Mir.

This was just after the Soviet Union had collapsed, Russia was opening up to the world and anxious to gain foreign currency. It was an amazing trip—we flew from Moscow to Leninsk, the service town for the Cosmodrome at Baikonur, and then by bus to the spaceport. This was nirvana for space enthusiasts; the location of Baikonur had been a closely guarded secret throughout the Cold War, yet here we were, a group of Americans, two Canadians, two Belgians and me, welcomed into the very heart of the now Russian space program.

We watched the Soyuz launch from 500 metres away, only an old post and wire fence separating the viewers, who included cosmonauts and generals, from the rocket. Being that close, the vibration from the launch seemed to go right through my body—I was sure that I could hear my teeth rattling.

There was a party that night to celebrate the launch, where the vodka flowed freely. I vaguely remember standing on a table with my arms around the two Canadians, singing their national anthem, O Canada. I swore off vodka that night and haven't touched a drop since.

The next day, we went to the launch pad of an unmanned Proton rocket, which was to be sent into orbit the next day with a Russian communications satellite. We were allowed to climb the ladder, which spiralled up the side of the rocket to inspect it. Imagine asking to do that at Cape Canaveral!

We then had permission to go to the pad from where the Soyuz had been launched the previous day. As we walked around, picking up pieces of burnt concrete as souvenirs, our guide showed us marks painted in the Cyrillic alphabet on the side of the launch gantry—the first one read 'Sputnik 1' and further down 'Vostok 1'. This was the very launch pad from which the first satellite and the first man in space were sent into orbit—truly hallowed ground for space enthusiasts.

Our three days at Baikonur ended with a visit, first, to a huge hangar where the second Russian space shuttle, Buran, was being built—the program was cancelled because of the cost to the struggling Russian economy. Then to the little house where Yuri Gagarin spent his last night before his historic flight in 1961, with the Chief Designer for the Soviet space program, Sergei Korolov.

The tour ended back in Moscow where, surrounded by more cosmonauts and generals, we watched on giant TV screens, the docking of the Soyuz with the Mir space station. It was a remarkable trip, which whetted my appetite to visit other launch sites around the world.

In late 1992, I got myself invited to the launch of an Optus communications satellite from Xichang in China, in 1993 I took some of the Americans and Canadians from the Russian trip to French Guiana, on the east coast of South America, for a European Ariane flight. In 1997, I was invited to a Japanese H-2 mission from the beautiful island of Tanegashima, carrying a weather satellite and scientific payload. 1994 saw me at Cape Canaveral for a Space Shuttle flight and I returned in 2006 for the launch of NASA's New Horizons spacecraft to Pluto.

Now there are further moves to have Australia become a space-faring nation. The Space Industry Association of Australia (SIAA) has released a White Paper advocating for a space agency and the Australian government has in July 2017 set up another inquiry. With the impetus of the city of Adelaide hosting the 2017 International Astronautical Congress, and with New Zealand having set up a space agency in 2016, the federal government may, at last, be convinced, or shamed, into action.

Whatever happens, I will continue to be fascinated by space research and development and by the enormous potential that it offers to Australia and to the world.

Peter Hammer

After Australis OSCAR 5 was launched, I was still very interested in amateur satellites, but the rest of the AO5 crew went off to do other things. With the support of AMSAT and the US Embassy in Canberra, I was receiving thick weekly packages of information and components, which probably didn't go down at all well with Prof Hopper. I went on to develop the command system for OSCAR 6. This was built using the very new first generation of integrated logic developed by Fairchild. It was simple RTL (resistor-transistor logic) and packaged in epoxy sealed ceramic disks. Not qualified for space use; however, it all worked perfectly. From here, the next project I got involved with was the Radio Teletype (RTTY) telemetry system for OSCAR 7. This was built using military grade CMOS logic[2] made by the RCA Corporation. Again, a brand new technology, which my lecturer in electrical engineering said would never replace transistors! He was totally wrong as it is the standard system used nowadays in every integrated circuit and image sensor. These chips were extremely sensitive to static electricity, which meant that special precautions had to be taken to make sure everything, including me, was properly earthed when working with these chips.

Many experts predicted that these circuits were totally unsuitable to space use and would be lucky if they lasted a few weeks in the harsh environment of space. Again, they were wrong as the RTTY telemetry system is still working after around 40 years in orbit and indeed the OSCAR 7 spacecraft holds the world record for the longest working satellite in Earth orbit. After the launch of OSCAR 7, I was pretty burned out and lost interest in amateur satellites and amateur radio and got involved with life and to a degree with my other hobby, photography. The boards for OSCAR 7 were made commercially through a company that was prepared to make them for me for free. All components had to be hand-soldered into place.

---

2   Complimentary Metal Oxide Semiconductor logic, now the mainstay of digital electronics.

After finally finishing my PhD at the end of 1972, I started work at LaTrobe University, where I was helping postgraduate students in the Physics Department.

For OSCARs 6 and 7, I was working at LaTrobe University and we had a very good 8" x10" camera in the electronics workshop. Technology had now moved to laying out the track pattern at twice full size using black precision cut tapes, which came in a range of widths. The tape was placed on a transparent plastic film, which was then photographed and reduced in scale to the correct size. This produced negatives on large scale film. Copper clad board was coated with a photosensitive resist (or you could purchase pre-coated board) in a darkroom lit with red light. The copper clad board was then sandwiched between the top and bottom layers of the track pattern and exposed to white light to burn the track pattern into the resist. The exposed boards were then etched in ferric chloride to dissolve the unwanted copper and the required holes were then drilled.

I married, had a son and eventually divorced. I have travelled considerably over the years and have now visited about forty countries and all seven continents. After a trip to Africa in 2005, I joined a camera club and started entering competitions, the Australian Nationals to begin with and later international salons. Since then, I have accumulated around 2,200 acceptances in international salons, 137 awards in international salons and a total of 17 gold medals. I am a Grand Master of the Australian Photographic Society. *(Note from the author: It is well worth a visit to his website: www.peterh.photography.)*

Steve Howard

After completing an electrical engineering degree, Steve joined the research laboratories of the Post Master General's Department, working first on microwave radio propagation in northern Australia, then on the design, construction and installation of solar radiometers to measure the attenuation of microwave radiation at 11 and 15 GHz in tropical conditions. Data from this work was used in the design of satellite communication systems.
In 1981, Stephen started his own company, Tain Electronics Pty Ltd[3], where he has developed a wide range of monitoring and control equipment for use in education, environmental research, agriculture and industrial applications. His latest product is a range of data loggers for monitoring soil moisture in the agricultural industry.

Rod Tucker

I completed my PhD in Microwave Engineering at the University of Melbourne in 1975 and then continued my research at the University of California in Berkeley, Cornell University, and the Plessey Company in the UK. In the late 1970s, I became fascinated by the emerging field of optical fibre communications, and joined Bell Labs in New Jersey to pursue a career in this area. I was located at Bell Laboratories, Crawford Hill, the location of some of the very early Telstar experiments. In the early 1990s, I came full circle, and returned to the University of Melbourne as a Professor of Electrical Engineering and Head of Department, where I worked until my retirement in 2014. In that time, I

---

3   See www.tain.com.au.

taught many courses and supervised approximately 40 PhD students. One of the highlights of my time at the University of Melbourne was my participation in a Panel of Experts that advised the Australian Government on the establishment of the National Broadband Network.

JAN KING

We've read that Jan King was a young engineer at NASA Goddard where he was participating in testing of all manner of NASA spacecraft, earth orbiting, lunar landing and interplanetary. He met many technical experts and, because of his involvement with AMSAT and Australis, he met many of the most senior managers within NASA. These meetings most definitely helped his career. As he says, it helped him 'see a bigger picture of the world . . . it quickly formed my career'.

The AMSAT organisation advanced 'tremendously' from its decision to fly Australis. AMSAT has been involved with over one hundred satellites and not just amateur radio satellites, but satellites built by universities and others. It has flown on rockets from every launching country, including the former foes of the US: the Soviet Union, now Russia.

Jan now lives in Australia and is still mesmerised by the excitement of space and spacecraft; he truly is an engineer of the first order. He continues to work on space projects and, as I write this, is in Svaalberg, Norway, where he is establishing a highly specialised receiving station for a series of satellites he is working on.

Reflecting about Australis and his life in Australia, Jan is less than complimentary about the risk culture in this country and the fact that we still have no space program of any kind, despite many attempts to interest various Australian governments. I like his referring to Canberra as 'Can't-berra'.

In an interview, Jan likens the engineering of spacecraft to the Earth itself. The creative effort to design a spacecraft to measure, or do something totally that has never been done before, has parallels to the challenges facing all the peoples of the spacecraft we call Earth. New ideas are needed to engineer new solutions to new problems on a new spacecraft and on spacecraft Earth. As we are finding with our young Australian engineers and people, they are no longer waiting, as we did after AO5, for governments to take a lead, but are just getting on with developing their ideas, hindered unfortunately by red tape. Get out of the way, "Can't-berra", and let these people flourish.

* * *

MY VW

The 'Vee-dub' served Australis and our early married life well. However, loading an infant into the back seat of a two-door car is difficult and so we traded it in for a larger, four-door car – still German, but this time second-hand.

CHAPTER 12
Advance Australia—Where?

Author's note: This chapter was written by Richard and it expresses our joint views. The chapter title is a reference to Australia's national anthem, 'Advance Australia Fair'.

* * *

Following the success of Australis, the question arose, 'What about Australis 2?' Why didn't the group go on to build another satellite, to start a business making small spacecraft, for a variety of uses? It would have created careers for the Project Australis team members, as well as many others, as later happened with Surrey Satellite, which grew out of enthusiasts like us, at Guildford University in Surrey, England. That company, now Surrey Satellite Technology Ltd, has grown to employ 485 commercial staff with annual revenues of £100 million and total annual export sales in excess of £600 million in 18 different countries. The founder, Martin Sweeting, has been knighted for his services to space technology.

But it was not to be. As Owen said in Kerrie Dougherty and Matthew Jones' excellent book, *Space Australia*:[1]

> We just couldn't face another one and we had dispersed and finished our degrees and went out and started learning what was not possible—what the real world looked like! . . . I guess in a way I regret now that, if I had been older and wiser, I would have started a business out of it because we had it all there. If we could have pulled it together as a space business, as Canada did for instance, then I think Australia would already have been one of the space faring nations of the world today, rather than aspiring once again to become one.

Tellingly, we did not even get a letter of congratulation from the Australian government for the Australis launch.

In order to understand why Australia is one of the only developed Western nations without even a space agency, a necessary basis for a country having even a modest space program (we are the only member of the United Nations Organisation for Economic Co-operation and Development, OECD, apart from Iceland, not to have an agency), it is necessary to look at the history of satellite construction and launching in Australia. It is a story of decades of frustration and failure, interspersed with brief moments of success and optimism.

In the late 1940s, in the aftermath of Britain's near defeat in World War II, the Woomera Rocket Range was established in outback South Australia, at the request of the United Kingdom government.

---

1   Kerrie Dougherty and Matthew Jones, Space Australia: the story of Australia's involvement in space' (Sydney: Powerhouse Publishing, 1993), 54. This book is being substantially rewritten and updated for the 2017 International Astronautical Congress to be held in September 2017 in Adelaide, Australia.

The history of the range, and of the Weapons Research Establishment (WRE) at Salisbury, north of Adelaide, is described in *Space Australia*.

In an early attempt at a European space program, ELDO, the European Launcher Development Organisation, was set up to build a satellite launching rocket to give Europe its own access to space, independent of the United States and the Soviet Union. Woomera was the obvious site to test the rocket, which appropriately was called Europa.

Europa was a mongrel—it had a British-built first stage, Blue Streak, which was a derivative of the American Thor missile, a French second stage, Coralie, a West German third stage, Astris, and an Italian-built test satellite. It was probably doomed from the start—the French hated the British, a feeling that was mutual, the Germans hated the French, and everybody hated the Italians.

Three attempts to launch test satellites with Europa from Woomera in 1968, 1969 and 1970 all failed to reach orbit—the last on 12 June 1970, six months after Australis' successful launch. Britain, in one of its many project cancellations of that time, withdrew from ELDO in 1968. The last Europa was launched from French Guiana in South America. It maintained the program's perfect failure record. To my knowledge, Australia's only real contribution to ELDO was to provide the dirt on which the launch facilities were built in the South Australian desert. The sad remains of ELDO can be seen in the rusting launch pad at Lake Hart.

Earlier, in 1967, Australia was given the opportunity to build our own satellite. In 1966, ten Redstone/Sparta rockets were shipped to Woomera from the United States in connection with atmospheric re-entry tests. As described earlier, Redstone was developed by Dr Werner von Braun's World War II V2 missile team as a short-range ballistic missile. It had the honour of launching Explorer 1, America's first satellite, and the first Mercury sub-orbital flights, piloted by Alan Shephard and Virgil Grissom. Solid rockets were added to the Redstone to drive the test vehicle into the atmosphere.

At the end of 1966, after nine flights, there was one Redstone/Sparta left over. This was offered to Australia to launch our own satellite, to be built by scientists and engineers at WRE. They, and the University of Adelaide in South Australia, already had experience in building payloads for sounding (that is, straight up and down, not into orbit) rockets. While everyone else involved was enthusiastic about the idea, the Federal government was not. They appeared unable to realise the scientific, technological and indeed the prestige benefits of the launch, and they initially refused.

Here the conspiracy theories start. Owen has already related the mysterious telephone call he received from a lady at WRE asking about the details of Australis and when it would be launched. That was a few months after we had taken Australis to California and we were awaiting a launch date—we thought before the end of 1967. The phone call showed that WRE, and therefore the government, knew that the scruffy University students had built the first Australian satellite and that its launch was approaching. Could that have changed the government's mind about approving WRESAT? We may never know, but we think it is a good story.

WRESAT was successfully launched into orbit from Woomera on 29 November 1967. WRE very kindly asked Project Australis (yes, they certainly knew about us) to send a delegation to watch the launch and, as Owen described, Peter missed the launch but I was there at Woomera when Australian space history was made. As a rather nice gesture, the day after Australis' launch in January 1970, WRE sent us a plate from their Baker-Nunn satellite tracking camera showing the ITOS-1 weather satellite and Australis, or Australis OSCAR 5 as it was then known, as streaks of light against the background of stars.

The authors of *Space Australia* said of the launch of WRESAT:

> The satellite success was a notable achievement for Australian scientists and engineers who hoped that its success would be the key to convincing the federal government to invest in a modest ongoing Australian space program . . . Unfortunately, although the satellite succeeded in its scientific mission, it failed to provide the hoped-for impetus for a commitment to space from the government or the development of a national space policy.

There was simply no government interest. It was rumoured at the time that there were plenty of Redstones on offer at the US Army's arsenal in Alabama, which could be made available for Australia's use to form the basis of a space program at little cost. The offer was refused.

WRESAT was not to be the last satellite launched from Woomera—not quite. On 28 October 1971, some years after WRESAT, a British Black Arrow two-stage rocket successfully put the British-built Prospero satellite into orbit. Of course, it had no impact on Australia, except that our desert was a safer and more politically correct launch site than the west coast of Cornwall.

The saga of Fedsat well illustrates the lack of a plan for a space industry. Fedsat was yet another *ad hoc* satellite project, a once-off launch and no follow on. The Co-operative Research Centres Program[2] (CRC) for Satellite Systems 'was established in January 1998 with an eight year grant'. It conducts research in areas such as satellite systems for communication, space science and computer technology.[3] Among the collaborators was the Institute for Telecommunications Research (ITR) at the University of South Australia where Owen served as a member of the Advisory Board of Management before and during the time of Fedsat construction. For more on the remarkable story of ITR, see the side box.

Part of the proposal for the CRC was that a satellite, Fedsat, would be built and launched during the year of celebrations marking the centenary of the Federation of Australia. Now for those overseas readers without a knowledge of Australian political history, 'Federation' was the moment when the six Australian colonies of Great Britain came together to form an independent Australia on 1 January 1901, after having obtained the permission of the British parliament at Westminster.

---

2   The CRC program is 'a competitive, merit based grant programme that supports industry-led and outcome-focused collaborative research partnerships between industry, researchers and the community'; see https://industry.gov.au/industry/IndustryInitiatives/IndustryResearchCollaboration/CRC/Pages/default.aspx and https://www.business.gov.au/assistance/cooperative-research-centres-program.

3   Encyclopedia of Australian Science, 'CRC for Satellite Systems. (1998-) at <http://trove.nla.gov.au/people/1471320?c=people>. Accessed 31 July 2017.

The Institute for Telecommunications Research (ITR) has had a stellar history in digital communications research, a discipline combining complex mathematics, electronics and computing. ITR's expertise was then in high-speed communications, demodulators for satellite ground stations, techniques for correcting errors in digital messages and it held the record for the closest approach to a theoretical limit for a communication channel (the Nyquist limit).

With this background, ITR became a member of the CRC for Satellite Systems funded by the Australian Government to build and launch Australia's first multi-experiment satellite – to be known as Fedsat – to celebrate our nation's Centenary of Federation year in 2001. ITR undertook the design and construction of an advanced communications experiment package for Fedsat. In addition, ITR's other challenge was to build and operate the telemetry, tracking and control Ground Station that would control the satellite in its orbits around the Earth.

Fedsat was successfully launched from Tanegashima in Japan on 14 December 2002. It continued to communicate with its UniSA Ground Station and permit operational experiments until August 2007 when it was turned off because its batteries were failing.

Australia hadn't built a satellite since WRESAT and Australis in 1967, over thirty years before Fedsat was thought of, so we were a little handicapped in spacecraft design and construction. Of course, we certainly could not launch it from Woomera—the last of the Redstones had decades before ended up in American museums and Prospero was the one and only satellite launched by the British Black Arrow as that program was cancelled.

The structure and services sub-systems of Fedsat were contracted to a British company, which went bankrupt partway through construction. To their great credit, a "Can't-berra" company, Auspace, literally picked up the pieces and finished the job. The Japanese space agency, NASDA, now JAXA, very kindly gave us a free launch as a secondary payload on their H2 rocket.

Nevertheless, the critical and innovative parts of the satellite, the electronics, were made in Australia, in particular the communications package designed and built by ITR; see the side box for more details.

So, finally, on 14 December 2002, Fedsat was successfully launched into orbit. The fact that it missed the centenary of Federation on 1 January 2001 by nearly two years was swept under the carpet. To be fair, it produced useful scientific data until May 2007 when its battery failed.

Now those criticisms would have been forgotten if the Australian government had grasped the opportunity, which they had failed to do after WRESAT, and used Fedsat as the basis for a modest scientific and applications space program. Alas, again, it was not to be. Following the announcement that Fedsat would be launched in five years' time, Richard wrote in a paper presented to a space conference in May 1997:

> The fact is that Fedsat is a stunt. It does not take five years to design and build a fifty kilogram satellite (in fact it took seven years). There are no funded plans for a follow on to Fedsat and, while I support the building and launching of an Australian satellite, the fact is that Fedsat looks like being another orphan, like WRESAT, with no presently planned programme to follow it. The launching of Fedsat in 2001 will simply emphasise the gap of thirty four years since our last "official" satellite, WRESAT.

My words were prophetic—there was no Fedsat 2. Indeed, it was not until 2017, fifteen years after Fedsat's launch, that Australian-built satellites would again circle the earth. Three cubesats, compact enough to hold in the palm of your hand, built by universities in Adelaide and Sydney and a European consortium, were launched from the International Space Station in May 2017.

In 1971, again in 1974, and more recently, the European Space Agency (ESA) has invited Australia to join them as an associate member, as Canada has, in recognition of our providing Woomera for the ELDO launches. Inexplicably, on each occasion the federal government declined, saying that it could not see any benefit in Australia being involved in space research! The ESA charter provides that contracts for the development of space systems are allocated to member countries in proportion to their contributions to the agency. ESA says that its members derive benefits of up to three times the amount that they invest in space activities. Australia, by its refusal to join ESA, has lost the opportunity to develop, over the past 46 years, a space industry for the country. These were fundamentally flawed and naive decisions, made by governments on both sides of politics, which at least demonstrates bipartisan ability to get it very wrong.

The list goes on—the Aussat programme was established in the early 1980s to provide a domestic communications satellite service for Australia from geostationary orbit[4]. Regrettably, less than one per cent of the three Aussat satellites and less than five per cent of the later Optus satellites were built in Australia. The same applies to the two recently commissioned National Broadband Network spacecraft that were built in the United States and launched on the European Ariane rocket from French Guiana. This was because Australian space-related industries had little experience in manufacturing satellite components as a direct result of the lack of substantial local space activity.

---

4   A geostationary orbit is one where the satellite's speed exactly matches the rotation of the earth and so appears to be stationary in the sky above the equator. The altitude of a geostationary orbit is 36,000 kilometres.

There is a postscript to the Aussat story. NASA offered to fly an Australian astronaut as a payload specialist on the Space Shuttle for both the Aussat A-1 and A-2 flights in the mid 1980s. Of course, the offer was declined. Thus was lost a unique opportunity for Australia to be given first-hand experience of human space flight, to forge closer links with NASA and to create widespread Australian public interest in space research and development.

Various organisations were set up by successive Australian governments over the decades to dabble in space research—the Australian Space Office, the Space Board, the Space Council, the Co-operative Research Centre for Satellite Systems. Some of those resulted in hardware being built, such as radiometers for European satellites and the Endeavour space telescope, which flew on the space shuttle. The University of Adelaide flew cosmic ray detectors on NASA's Pioneer 6 and 7 deep space probes, but there was never any follow on, no long-range plan that would have allowed an Australian space industry to grow and thrive[5].

In the mid 1990s, I was the Deputy Chair of the National Space Society of Australia and a member of the Executive Council of the Australian Space Industry Chamber of Commerce (ASICC). ASICC developed a plan for an Australian National Space Agency (ANSA) and presented it to the Labour government and the coalition opposition ahead of the 1996 federal election. While the plan was well received, and it may have helped in the government's decision to go ahead with Fedsat, the space agency never materialised.

But does Australia really need a space agency and a modest space program? The facts and figures are overwhelming. Apart from being almost the only developed nation not to be actively involved in space activities (New Zealand formed a space agency in 2015 and is now host to Rocket Labs, an American company launching satellites from the North Island), we are spending in excess of one billion dollars each year on space services and related activities when we could make and supply much of that in Australia. Those purchases contribute a significant amount to our balance of payments deficit while adding little to our technological or economic base. With an appropriate space program, many of those expenditures can be left inside Australia, which would not only decrease imports but materially assist in the expansion of our electronics, software and other space-related industries. The cost of 'pump priming' an Australian space industry would be a fraction of that one-billion-dollar expenditure. Add to that the opportunities that are being lost for space science, engineering and other graduates from our tertiary institutions who must go to live and work overseas if they are to pursue careers in space research and development.

Space-related systems affect us all in our daily lives. Your mobile phone, the sat. nav. in your car, bank Automatic Teller Machines, telephone and television, weather forecasting, farming, resource management and a myriad of other services rely on satellites for their successful operation.

---

5   Owen's comment: there are close parallels in the defence industry, especially naval ship building.

In conclusion, Australia has squandered many opportunities over the fifty years since WRESAT and Australis to establish a viable, economically productive space industry to serve the nation. As a result, we are almost alone among the developed countries of the world in not having our own space program.

It will take years and a commitment of funds for the decades of neglect to be made up. However, there is no doubt that such a commitment is worthwhile and that it will result in a vigorous, innovative, high technology industry for Australia, which will lead to significant exports of space hardware, software and services within a decade.

While most Australians believe in space research and the benefits that it can bring to the country, the supporters of an Australian space program have always had a hard road to follow. The task has been difficult and there have been few rewards over the past fifty years. However, history shows that those who have a vision, who have a goal to work toward and who apply themselves diligently to the task, despite the many disappointments along the way, usually achieve that goal in the end, as did those scruffy nerds from the University of Melbourne who built Australis all those decades ago.

* * *

Owen's comments: Australia does not seem to possess the 'can do' of the American of the era of this book. I believe the lack of government interest in space is part of a wider community belief and culture that Australian technology and designers are unable to compete with overseas experts. This is, of course, a self-fulfilling belief—Australians can't do it, so we'll give it to a foreign outfit. There are numerous examples and no doubt there's another book in it, too. It is also incorrect. As many will tell you, Australian engineers and scientists are among the most innovative in the world—we have to be to compete in a world of highly specialised people – and we tend to be generalists and perhaps see a broader view of an engineering challenge. Australians punch above our weight in Nobel prizes and the like.

CHAPTER 13

Postscript

If I think about the outcomes, two become clear: first, the effects on our lives, which were significant at least. Both Richard and I found our life partners as a direct result of our participation in Australis; see the 'Pauline Remembers' side box. Peter went on to achieve spectacular success with his work on OSCARs 6 and 7. Steve, Paul, Rod, Dave and the others also gained immeasurably from their involvement.

Second, I record with pride the firsts that we achieved with AO5:

- First Australian-built satellite (WRESAT was built after AO5 but flown before).

- First non-US amateur satellite.

- First University satellite.

- First non-government secondary payload and the first amateur radio satellite launched by NASA.

- First satellite, along with Tiros M, to be launched by the then-new Delta N rocket.

- First 10 metre HF transmitter flown on an amateur satellite.

- First command system flown on an amateur satellite.

- First 'PMASS' magnetic stabilisation and orientation system to stabilise and orient a satellite. There have been about 40 subsequent systems flown.

- First use of 'Standard Orbits' method for predicting passes and for tracking. I was told that the method was later adopted and improved by NASA for use by US Antarctic expeditions.

- First time that satellite predictions (in this case equator crossings) had been transmitted directly from a computer to users via radio without human intervention.

In addition, AO5 trained two groups, the Australians and the US AMSAT group, whose achievements have been noted.

At the 23 July 2011 reunion of the group, Peter's contribution to AMSAT with his command receivers was acknowledged: the AMSAT 7 command system continues to operate, along with the RTTY telemetry, after some 40 years since launch, and it is the longest surviving space command system ever.

It was to over eleven years before the Surrey University group under Professor Martin Sweeting[1] flew the first of their small and relatively inexpensive satellites.

As it turned out, we were pioneers, but not intentionally so. We were simply following our interests.

May the young generations of engineers and scientists do the same, for they are the real leaders of our technologically based society.

---

1   University of Surrey, 'Professor Sir Martin Sweeney' at <www.surrey.ac.uk/ssc/people/martin_sweeting>. Accessed 31 July 2017.

PAULINE REMEMBERS

I first met Richard at the house in which he and some friends lived in Drummond Street, Carlton. As Owen has described, I was invited as a 'visiting cook' to prepare and cook the evening meal for Richard and his friends. Bizarre though that may seem today, it was actually an honour for a young lady to be invited to Drummond Street, Carlton, not only to cook the evening meal, but to bring the meal with her.

Owen's version of the 'satellite in the oven' is correct—there really were metal boxes with wires coming out of them inside the 'Kookaburra' gas oven in the kitchen of the terrace house in Carlton. And Richard really did delay me cooking the meal until he had satisfied himself that he had completed the tests on the satellite parts.

About two months later, after Richard, Owen and Paul had taken the satellite to California and after I had returned from a Melbourne Symphony Orchestra tour to the United States and Canada, Richard arrived at one of the parties that my housemates and I threw. It was there that I noticed Richard, properly, for the first time—this nerdy but terribly handsome young man with Buddy Holly glasses. I experienced several terrifying rides on the back of Richard's motor scooter before he finally bought a car, a grey Toyota Corolla, in about mid-1968. Our romance flourished and we were married in January 1969, almost exactly a year before Australis was launched.

I supported Richard during the long wait to find out whether Australis was going to be launched or returned to Australia. However, on 23 January 1970, I joined the others in the MUAS control room at Melbourne University to listen to the successful launch of Australis. At last, we could all get on with our lives and our first child, Karen, was born in the middle of 1971.

Glossary

ABLS	Australian Balloon Launch Service
AO5	Australis Oscar 5
APT	Automatic Picture Transmission
ASIO	Australian Secret Intelligence Agency
Azimuth	The angle, measured clockwise along the horizon, from north to a point on the horizon[1]. See also Elevation.
BOM	Bureau of Meteorology, Australia
Case	Case Western Reserve University, Cleveland, Ohio
CIA	US Central Intelligence Agency
CRC	Co-operative Research Centres Program
Elevation	The angle measured from the horizon to a point above (or below) it. See also Azimuth.
FCC	Federal Communications Commission
GeV	Giga-electron Volts; that is, billions of electron volts
GPS	Global Positioning System
HF	High Frequency radio waves 3 MHz to 30 MHz (wavelengths from 100m to 10m)
IBM	International Business Machines company
IC	Integrated Circuits; that is, microelectronics
IGY	International Geophysical Year, June 1957 to December 1958
ITOS	Improved TIROS (q.v.) Operational Satellite
ITR	Institute for Telecommunications Research at the University of South Australia
JIB	Joint Intelligence Bureau, Melbourne, Australia
JPL	Jet Propulsion Laboratory, California
LSD	Lysergic Acid Diethylamide, a hallucinogenic drug favoured by hippies in the 1960s

Glossary continued

MASS	Magnetic Attitude Stabilisation System used on Australis
MUAS	Melbourne University Astronautical Society
NACA	National Advisory Committee for Aeronautics the precursor to NASA
NASA	National Aeronautics and Space Administration founded in 1958 assuming the responsibilities of NACA
NOAA	National Oceanographic and Atmospheric Administration
NSBF	National Scientific Balloon Facility
OSCAR	Orbiting Satellite Carrying Amateur Radio
PhD	Doctor of Philosophy
PMHG	Australian Post Master General's Department
RAAF	Royal Australian Air Force
RTL	Resistor Transistor Logic
RTTY	Radio Teletype
SLAC	Stanford Linear Accelerator Center
STEM	Science, Technology, Engineering and Mathematics
TIROS	Television and Infrared Observation Satellite
VHF	Very High Frequency radio waves 30 MHz to 300 MHz (wavelength from 10m to 1m)
VW	Volkswagen
WRE	Weapons Research Establishment
WRESAT	WRE Satellite launched on 29 November 1967

1   https://en.wiki.org/wiki/Azimuth for a more complete definition

Users Guide

The Australis OSCAR A Users' Guide is reproduced in full and as it was published in 1969, beginning with the cover page and complete with "bleed through" from the reverse side of each diagram page:

AUSTRALIS OSCAR A

USERS' GUIDE

* * * * * *

The Australis OSCAR A Users' Guide contains full instructions for all wishing to track the satellite. Since the success of the project depends on the support of a large number of tracking stations,we are anxious to enlist the co-operation of suitably equipped radio operators, short-wave listeners and VHF enthusiasts everywhere. Further copies of this booklet are available from Regional Directors, or from:

> Project Australis,
>
> Union House,
>
> University of Melbourne,
>
> Parkville, Victoria,
>
> Australia. 3052.

Any enquiries or requests for more detailed information will also be welcomed by Project Australis.

CONTENTS

Frontispiece

Above: The completed satellite. The photograph shows the thermal
          control pattern on the metal surface.

Below: A top view of the internal compartment. The battery
          occupies the centre, and the electronics modules are
          mounted on bulkheads at each side. The ends of the VHF
          transmitting antenna, and the receiving antenna are
          visible.

1. BACKGROUND

The Melbourne University Astronautical Society was formed at a time when the image of space research was dominated by a spirit of adventure. Today, much of the popular interest has subsided, but the potential of the spacecraft is being rapidly revealed. The satellite is an indispensible tool in many fields of research; its use in communications, navigation and meteorology is commonplace. The matter of communications, which received major publicity in 1962 with the success of Telstar 1, had already attracted the attention of amateur radio operators in the U.S.A.

At present the HF bands are overcrowded, but the traffic increases daily. One obvious solution is to move to higher frequencies. The early problems of noise and instability no longer haunt the VHF bands, but propagation characteristics severely limit the capabilities of VHF. Global communications may be achieved by such methods as moonbounce, but perhaps another solution is the artificial satellite. This has been accomplished, but still the amateurs are tied to the HF bands for international communications.

During 1965, the Melbourne University Astronautical Society began to investigate the problems of satellite construction. With the co-operation of Oscar, Project Australis was formed. Australis, like Oscar, aims to build communications satellites for use by amateur operators in all parts of theworld. In contrast to its American counterpart,Australis has no local background of satellite technology. This situation contributed to the difficulty in initiating the project. Financial limitations have also restricted progress. The result is that the first satellite is a relatively simple test vehicle, carrying two telemetry transmitters, a command system and a magnetic attitude control system. All electrical power is supplied by batteries which are expected to have an operating lifetime of about two months.

The satellite does not carry a repeater or translator.

It will be known as Australis OSCAR A until it is placed in orbit around the earth. Once in orbit, it will be given the next number in the OSCAR series to replace the "A".

The package construction, the command system, the antenna array and the magnetic attitude stabilization system could all be classed as experimental. The rest of the satellite provides the platform on which the experiments may be conducted.

However,when the experimental data must be recorded at a distance, the techniques of information transmission are added variables in the system.

For amateur operators and short-wave listeners there are opportunities to practise the art of tracking satellite signals in both the ten metre and two metrebands. The behaviour of the ten metre signal will illustrate long range propagation characteristics in the band.

In addition, there is a secondary objective. The project requires an efficient ground communications system to disseminate orbital figures and to collect data recorded by operators in all parts of the world. So far, the information channels have been organised,but the reliability of such a system has yet to be proved.

The final point illustrates the dependence of the project on human,as well as technical factors. Mechanical strength may be measured; electronic reliability has been improved with technology; for your assistance and co-operation we can only ask.

2. A TECHNICAL DESCRIPTION

2.1 Introduction.

The electronics of the satellite may be represented by the block diagram of figure 2.1. The physical layout is shown infigure 2.2.The obvious essentials are the two transmitters (10 metres and 2 metres) carrying the eight channel telemetry. To conserve battery power,a command system allows ground stations to control the operating time of the H.F. transmitter. A timetable will be published before the launch. A brief technical description of the spacecraft follows.

2.2 Hi Keyer

The hi keyer generates the Morse code identification. Although it operates continuously, producing the synchronization pulses for the telemetry encoder,its signal is transmitted for only 6½ seconds of each telemetry cycle.

2.3 Telemetry

Temperature, spin rate and battery performance are relayed to earth by the eight channel telemetry. Two temperature readings - one at the inside surface of the aluminium case, and the other from the insulated electronics compartment - are effected by thermistors.

Three phototransistors sensitive to reflected radiation from the earth are mounted on orthogonal axes. The output from each will indicate its orientation, and the rate of variation of all three is a measure of spin velocity. The channel sequence is:

0	hi identification
1	current drain
2	X axis horizon sensor
3	battery voltage
4	Yaxis horizon sensor
5	internal temperature
6	Z axis horizon sensor
7	skin temperature

The X, Y, Z,axes are as defined in figure 2.2 and figure2.9.

In every case, the parameter is specified simply by the audio frequency. Unlike OSCAR 1 and OSCAR 2, the hi channel carries no telemetry data.

FLIGHT CONFIGURATION

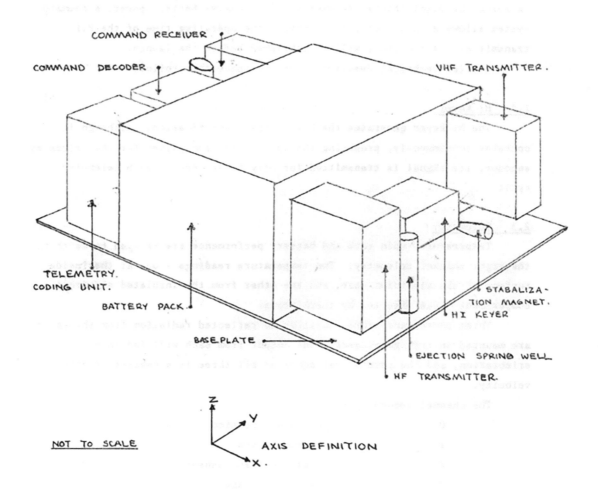

COMMAND RECEIVER

COMMAND DECODER

VHF TRANSMITTER.

TELEMETRY.
CODING UNIT.

BATTERY PACK.

BASEPLATE

STABALIZA-
TION MAGNET.

HI KEYER

EJECTION SPRING WELL

HF TRANSMITTER.

NOT TO SCALE

Z

Y

X

AXIS DEFINITION

AUSTRALIS OSCAR 'A' SATELLITE

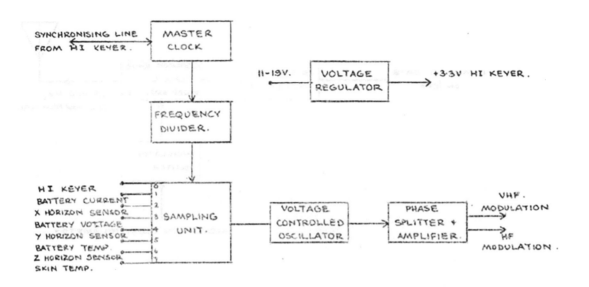

SYNCHRONISING LINE
FROM HI KEYER.

MASTER CLOCK

11-19V. → VOLTAGE REGULATOR → +3.3V HI KEYER.

FREQUENCY DIVIDER.

HI KEYER
BATTERY CURRENT
X HORIZON SENSOR
BATTERY VOLTAGE
Y HORIZON SENSOR
BATTERY TEMP.
Z HORIZON SENSOR
SKIN TEMP.

0
1
2
3
4
5
6
7

SAMPLING UNIT.

VOLTAGE CONTROLLED OSCILLATOR

PHASE SPLITTER + AMPLIFIER.

VHF. MODULATION

HF MODULATION.

FIG 2·3   TELEMETRY ENCODER BLOCK DIAGRAM

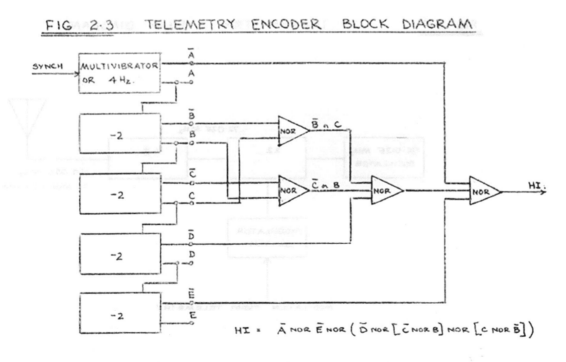

SYNCH → MULTIVIBRATOR OR 4Hz.

$\bar{A}$
A

−2

$\bar{B}$
B

−2

$\bar{C}$
C

−2

$\bar{D}$
D

−2

$\bar{E}$
E

NOR $\bar{B} \cap C$

NOR $\bar{C} \cap B$

NOR

NOR   HI.

HI = $\bar{A}$ NOR $\bar{E}$ NOR ($\bar{D}$ NOR [$\bar{C}$ NOR B] NOR [C NOR $\bar{B}$])

FIG 2·4   HI KEYER   BLOCK DIAGRAM

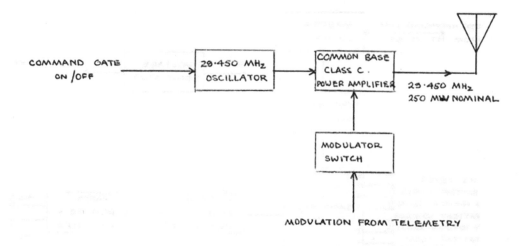

FIG 2·5   HF TRANSMITTER   BLOCK DIAGRAM

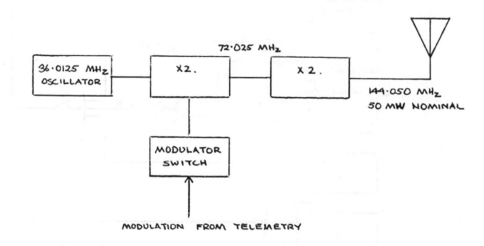

FIG 2·6   VHF TRANSMITTER   BLOCK DIAGRAM

A continuously operating switch ("encoder") samples each sensor for about 6½ seconds in the 52 second cycle. The voltage output is fed to an audio oscillator which modulates both transmitters. The audio frequency may vary from 400 Hz up to about 2,000 Hz. Graphs relating frequency to individual parameter values are given in chapter four.

2.4 V.H.F. Transmitter

A 50 mW crystal controlled transmitter operates continuously on 144.050 MHz. It is amplitude modulated by the telemetry.

2.5 H.F. Transmitter

The only ground commandable equipment is the 250 mW H.F. transmitter. It is crystal controlled on 29.450 MHz. The modulationis identical with the V.H.F. signal, except for a 180 degree phase difference. In each case the modulation index is 0.90.

2.6 Command System

Commands from earth are detected by a double change superhet receiver. The audio output is fed to the decoder which determines the validity of the command. When a correct signal is received, the decoder produces a control voltage to switch the H.F. transmitter.

2.7 Battery

Power is supplied by 28 alkaline manganese cells wired in two identical 20 volt series "strings". Each string supplies one transmitter, and the rest of the electronics run from both strings through an arrangement of protective diodes. If one string fails by short circuit or open circuit, then one transmitter is cut out, but the rest of the system operates. The diodes ensure that a short circuit in one string cannot impose an excessive load on the other.

2.8 Stabilization

To limit signal fading, and to maintain the antennas in a favourable orientation, some form of attitude control is necessary. Spin may be introduced at ejection,or by the prolonged action during thesatellite lifetime, of microscopic perturbing torques. The energy associated with spin is removed by magnetic hysteresis loss in an array of permalloy wires, and by eddy current loss in the aluminium alloy case. A bar magnet brings the X axis of the satellite into line with the earth's magnetic field.

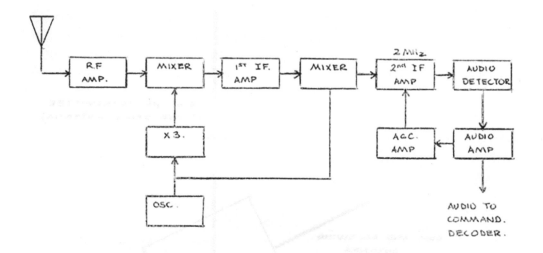

FIG 2·7   COMMAND RECEIVER BLOCK DIAGRAM

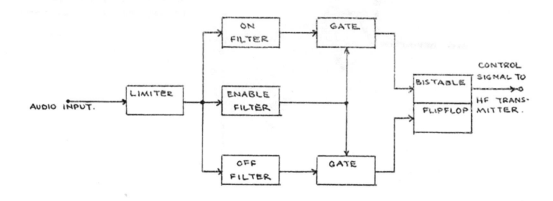

FIG 2·8   COMMAND DECODER BLOCK DIAGRAM

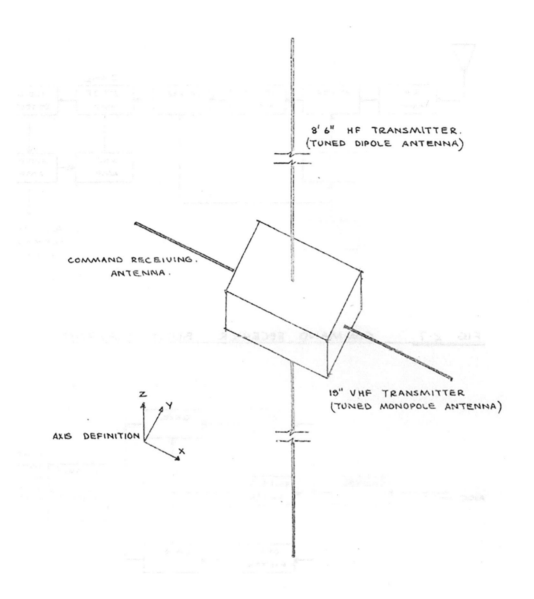

8' 6" HF TRANSMITTER.
(TUNED DIPOLE ANTENNA)

COMMAND RECEIVING.
ANTENNA.

15" VHF TRANSMITTER
(TUNED MONOPOLE ANTENNA)

AXIS DEFINITION

FIG 2.9 AUSTRALIS OSCAR 'A' FLIGHT CON-
-FIGURATION

2.9 Package

The electronics modules are mounted on an aluminium frame which is built around the battery compartment. A layer of thermal insulation separates all of this from the outercase. The aluminium alloy used for the case contains 1.0% magnesium, 0.6% silicon, 0.2% copper and 0.2% chromium. A paint pattern on the outside surface is designed to maintain a fairly stable internal temperature by regulating heat radiation.

All antennas are made of flexible steel tape, and are mounted as shown in figure 2.9.

3. TRACKING INFORMATION FOR AUSTRALIS-OSCAR A

3.1 Regional Directors

For the purposes of disseminating tracking information, three regional directors have been appointed. Each director is responsible for distributing information within a specified area. When Australis OSCAR A has been launched, Project Oscar will obtain orbital data and distribute them to the regional directors who will send them to localco-ordinators. Local co-ordinators will complete the distribution to all tracking stations within their area.

Table 3.1 Areas and Regional Directors

North and South America	Project Oscar Inc., Foothill College, Los Altos Hills, California, U.S.A. 94022.
Asia and Australasia	Project Australis, Union House, University of Melbourne, Parkvilie, Victoria, Australia. 3052.
Europe and Africa	W. Browning, G2A0X, 47 Brampton Grove, Hendon, London, N.W.4, U.K.

3.1.1 Data Distribution Within Asia and Australasia

The local co-ordinators within the Asian and Australasian area act as links between the regional director and amateurs who are tracking Australis OSCAR A. The co-ordinator will have the following responsibilities:

(a) He will have equipment to provide two-way h.f. communication with the regional director for the reception of tracking information and the transmission of urgent data about the satellite condition.

(b) He will distribute orbital predictions to amateurs within his area,

(c) He will provide telemetry forms to tracking stations and return completed forms to:

> Project Australis (Telemetry),
> UnionHouse,
> University of Melbourne,
> Parkville, Victoria,
> Australia. 3052.

(d) The Users' Guide may be released to the press.

Table 3.2 lists the local co-ordinators for the Asian and Australasia

Table 3.2 Local Co-ordinators for Asia and Australasia

New South Wales	A. Swinton, VK2AAK, P.O. Box 1, Kulnura, N.S.W. 2251.
Victoria	W.M. Rice, VK3ABP, 34 Maidstone Street, Altona, Victoria. 3018.
Queensland	L. Blagborough, VK4ZGL, 54 Bishop Street, St.Lucia, Queensland. 4067.

South Australia	B. Tideman, VK5TN, 33 Ningana Avenue, Kingspark, South Australia. 5034.
Western Australia	D. Graham, VK6HK, 42 Purdon Street, Wembley, Western Australia. 6019
Tasmania	P. Frith, VK7PF, 181 Punchbowl Road, Launceston, Tasmania. 7250.
Japan	Kenso Sano,JA1EC, 11-16 Misaki -2, Kofu, Japan.
Malaysia	C.W.C. Richards, 9M2CR, Telecommunications Training Centre, Jalan Gurney, Kuala Lumpur, Malaysia Cont'd.
New Zealand	B. Rowlings, ZL1WB, Mason Street, Onerahi, Whangerei, Northland, New Zealand.

3.3 Orbital Data and Predictions

3.3.1 Introduction

In order to obtain good v.h.f. telemetry records from Australis OSCAR A, it will be necessary to use moderately directive receiving antennas which must be pointed towards the satellite throughout the pass. This section describes the tracking data to be distributed by Project Australis and explain show to use it.

3.3.2 Using the Orbital Predictions

Throughout this section it is assumed that the satellite is in a circular orbit at a height of 500 statute miles, and with an inclination of 70 degrees to the equator.

Once the height and inclination of the orbit are known, the position of the satellite during a particular pass can be specified by the time and longitude of the previous north bound equator crossing of the satellite. The times and longitude of these northbound equator crossings will be predicted by Project OSCAR and distributed to local co-ordinators. A typical set of northbound equator crossings is given in table 3.3.

Each local co-ordinator will be provided with a set of standard antenna pointing angles, giving at two minute intervals, th esatellite azimuth and elevation angles and the number of minutes since the previous northbound equator crossing. These pointing angles will be supplied for a number of standard longitudes of the northbound equator crossing. A set of pointing angles for a standard pass is shown in table 3.4.

To obtain antenna pointing angles for a particular pass, choose the standard set which has a northbound equator crossing as close as possible to the actual longitude of the northbound equator crossing for the pass. This actual longitude will be given in the orbital predictions, such as those in table 3.3. Add the number of minutes given in the left-hand column of the set of standard pointing angles (table 3.4) to the time of the northbound equator crossing for the actual pass (given in the predictions, such as in table 3.3),obtaining the time for which the satellite is at the given azimuth and elevation angles.

```
ASCENDING NODES FOR AUSTRALIS OSCAR A


 DATE      ORBIT    TIME    WEST LONGITUDE

31JAN66    0000     0526       356
31JAN66    0001     0707        20
31JAN66    0002     0848        44    ***********************
31JAN66    0003     1029        70
31JAN66    0004     1210        96


              TABLE 3.3




              STANDARD ORBIT CORDINATES

  FOR STATION VK3ATM  MELBOURNE, AUSTRALIA,
      215DEG WEST,   37DEG SOUTH.

              ASCENDING NODE 45DEG WEST

       ADD MINUTES    AZIMUTH  ELEVATION

            84          171         3
            86          165         9
            88          159        15
            90          144        19
            92          131        15
            94          123        10
            96          119         5


              TABLE 3.4



  THIS IS A SAMPLE COMPUTER OUTPUT
```

For example, if orbit number 0002 of table 3.3 is to be tracked at Melbourne, first obtain the longitude of the north bound equator crossing from table3,3 (44 W). Then choose the closest standard orbit, for which the longitude of the northbound equator crossing is 45W (shown in table 3.4). To give the actual time, add the equator crossing time to each time in the left hand column of table 3.4. Thus at 0848 GMT + 84 minutes = 1012 GMT the satellite azimuth will be 171 deg. and elevation will be 3 deg. The azimuth and elevation angles are similarly calculated every two minutes, giving the pointing angles shown in table 3.5.

Table 3.5 Calculated Pointing Angles for Orbit Number 0002

Time GMT	Azimuth deg.	Elevation deg.
0848+84 = 1012	171	3
+86 = 1014	165	9
+88 = 1016	159	15
+90 = 1018	144	19
+92 = 1020	131	15
+94 = 1022	123	10
+96 = 1024	119	5

As a rule, tracking stations will be able to observe two northbound passes about 100 minutes apart,followed about 12hours later by two south-bound passes about 100 minutes apart. This pattern will be repeated each day.

3.3.3 Schedules

As a rough guide, the equator crossing predictions are accurate for as long after issue as the satellite has been in orbit when the predictions are issued. For example, predictions issued three weeks after launch will be accurate for about another three weeks.

Each local co-ordinator will receive tables of Standard Pointing Angles and Northbound Equator Crossings as described below.

    (a) Several months before launch,a set of Standard Pointing Angles for the expected orbit, and a set of typical Northbound Equator Crossings (for demonstration purposes only) will be issued.

    (b) As soon as possible after the launch, a list of Northbound

Equator Crossings will be issued. This list will probably be accurate for only a few days. If the actual orbit is greatly different from that expected, a new table of Standard Pointing Angles will be issued.

(c) Throughout the satellite lifetime, lists of Northbound Equator crossings will be issued by both mail and amateur radio, sufficiently often to keep local co-ordinators well informed, probably at fortnightly intervals.

4. USING AUSTRALIS-OSCAR A

4.1 Introduction

Australia OSCAR A will transmit telemetry continuously at a frequency of 144.050 MHz, and at a frequency of 29.450 MHz when the transmitter has been commanded on. All tracking stations are requested to obtain telemetry data from either transmitter whenever possible, since telemetry reception and reduction is one of the major purposes of this project. The following sections give an outline of the minimume quipment needed to receive telemetry from Australis OSCAR A.

4.2 Receiving Antennas

4.2.1 V.H.F. Antenna

It is desirable to use a circularly polarised receiving antenna to reduce fading caused by changes in satellite attitude. This antenna should have a gain of at least 10 db. One suitable antenna is a crossed Yagi (two Yagi antennas pointing in the same direction, one with vertical and the other with horizontal polarization), one being connected through an extra quarter wavelength of cable, giving a 90 degree phase shift between the two driven elements. Another suitable antenna is a helix, such as the one described in "QST" for November, 1965.

To receive good signals while the satellite is at high elevations the antenna should be steerable in elevation as well as in azimuth. If measurements of the satellite spin rate are to be made, a horizontally or vertically polarized antenna should be used.

4.2.2 H.F. Antenna

If a linearly polarized antenna is used to receive the h.f. signal, fading will occur because of both satellite spin, and ionospheric Faraday rotation. Thus it may be difficult to determine the satellite spin using the h.f. signal, unless the operator is capable of separating the two variations, For reception of the h.f. telemetry, a pair of crossed horizontal dipoles, mounted one quarter wavelength above ground, will give a reasonably good omni-directional,circularly polarized pattern.

4.3 Converters

To obtain a good signal to noise ratio, the h.f. converter should have a noise figure of about 4 to 8 db. Most h.f. receivers should be adequate to receive the h.f. telemetry although some older receivers may need a pre-amplifier.

4.4 Receivers

Both transmitters are amplitude modulated, with maximum modulation frequencies of 2,000 Hz, so that receivers should have i.f. bandwidths of about 4,000 Hz. Except for initial acquisition of the signal, a b.f.o. should not be used, as the telemetry information will be lost.

4.5 Telemetry

Most of the information required about the satellite is derived from the audio telemetry, which has eight sequential channels. Each channel is transmitted for about 6½ seconds and the whole cycle lasts for 52 seconds. The hi channel consists of a 1.6 sec. tone, followed by a 1.6 sec. hi all repeated once again. The hi is transmitted not as MCW but as AFSK. Thus the tones do not key on and off, but switch between two tones of different frequency. The actual frequencies contain no telemetry information.

The hi channel is followed by seven tones, each 6½ sec, long and each sending information about one of the channels. By measuring the audio frequency and using the calibration graph for the channel, the quantity concerned can be determined. During telemetry decoding, the time should be watched carefully, as the frequencies of two adjacent channels may be similar and the transition from one to the next may not be audible.

The sequence of the telemetry channels is given in table 4.1.

Table 4.1 Telemetry Channels

Channel Number	Function
0	hi in Morse code
1	total battery current drain
2	X axis horizon sensor
3	battery voltage
4	Y axis horizon sensor
5	internal temperature
6	Z axis attitude sensor
7	skin temperature

Calibration charts for decoding the telemetry are given for battery current (figure 4.1), battery voltage (figure 4.2), and for internal and skin temperatures (figure 4.3).

To enable the telemetry reports to be evaluated by computer, all tracking stations are requested to enter their observations on a special telemetry coding form, as described in section 4.8.

4.6 Telemetry Decoding

One convenient method for decoding the telemetry is to use Lissajous figures. The received audio signal is applied to the vertical input of an oscilloscope and a sine wave from a calibrated audio oscillator is applied to the horizontal input. The frequency of the audio oscillator is adjusted until a stationary ellipse is seen, indicating that both frequencies are the same.

If the oscilloscope timebase has been calibrated, a set number of cycles can be displayed and the period of each cycle determined, and hence the frequency. If the timebase is free-running, as little syncas possible should be used to avoid changing the timebase calibration.

If an oscilloscope is not available, the frequencies of the received telemetry and of the audio oscillator can be matched by ear. Even with poor signal to noise ratios this method gives results accurate to within about 10 Hz at 2,000 Hz.

Another method, which in many cases can give better accuracy than any previously described, is to match the tone with a piano note. However, confusion of octaves must be carefully avoided.

Lastly, if the signal to noise ratio is good, the best method is to use a direct-reading frequency meter or digital counter.

If a tape recorder is used to record data, its speed should be accurate to within five percent, at worst, or else results will be seriously in error. Otherwise, operators are advised to practise measuring the frequency of an audio tone in less than seven seconds. It should be pointed out that inaccurate results are worse than none at all - an accuracy of at least ten percent is needed.

4.7 Readability and Signal Strength

The readability and strength of the received signal will be used in deciding the weight given to the decoded telemetry.

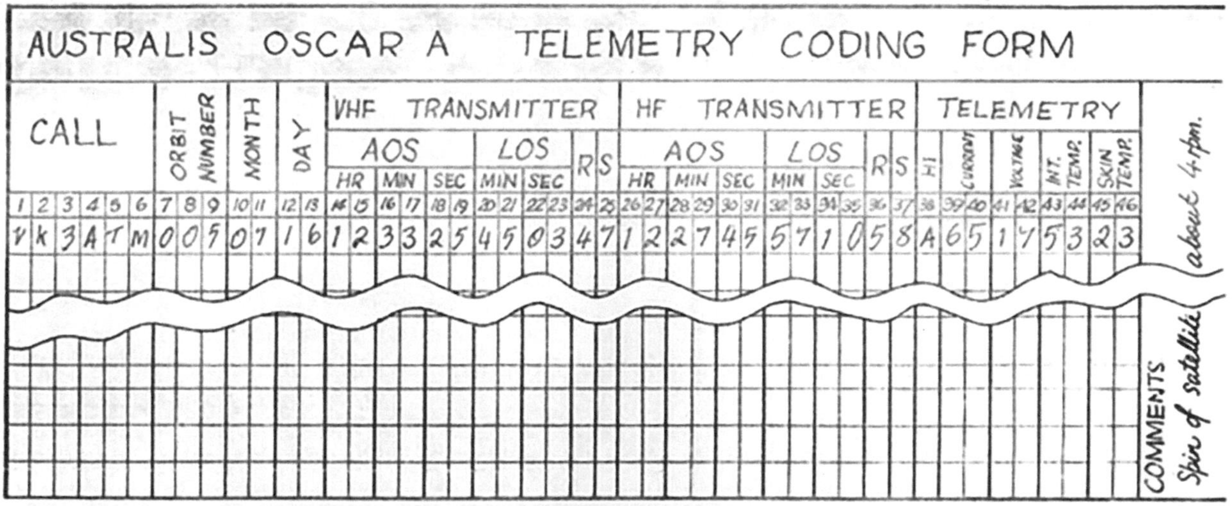

FIG 4.4 A TYPICAL PASS CODED - ALL SYSTEMS OPERATING NORMALLY

4.8 Telemetry Coding Form

Having decoded the telemetry for a pass, please select those results which you think are the most reliable. This will often mean rejecting wildly inconsistent results which may arise when the telemetry is decoded directly, rather than from a recording. Where a large,number of consistent results are obtained,all should be entered on the telemetry coding form, since this is an ideal indication of the reliability of the information.

Please write clearly, with only one character in each column.

All dates and times must be in GMT.

The following information is required :

(a) call sign of tracking station (if no call sign, write ZZ1, followed by the operator's initials)

(b) orbit number

(c) month and day

(d) time of acquisition of signal (AOS) and loss of signal (LOS), and readability and strength for each transmitter.

(e) hi keyer operation; the letter A for normal and F for failure, which should be described on a separate sheet.

(f) battery current drain in milliamp from figure 4.1

(g) battery voltage in volt from figure 4.2

(h) internal temperature in degree C from figure 4.3

(i) skin temperature in degree C from figure 4.3

All data entered on these sheets will be stored in a computer at Melbourne University. The form is in fact a replica of a computer card.

Reports on horizon sensor data should be treated differently. Since we are concerned only with either "light" or"dark",the actual frequency of the sound is of no interest. Each change in frequency corresponds to a transition of the field of view of a sensor between different states of illumination. The length of the higher frequency (bright) periods, depends on the spin rate,and on the nature of the traverse across the bright source. For example a short period could correspond to a single sweep a crossa short chord,or to a much faster sweep across a near diameter of the earth's disc. The sun and moon will also appear as bright sources against the dark background of space. However, they subtend such small solid angles at the satellite that the sensors will rarely sweep across them. Both would produce short high-pitched signals in the appropriate telemetry channels (Nos 2,4, or 6).

Now because the package may be rotating about three axes simultaneously, the spin rate on any single channel may not sound regular, except over a very long time. It is impossible to determine the spin rate directly. In fact it is a job for a computer, but this would require the recording of several telemetry cycles at various times. As far as individual operators are concerned, we would only expect a comment on the length of the sweeps across the earth. In this case, "fast" might be about one second; four seconds would be "slow", An average statement for each of the three axes is necessary.

Since computers do not take kindly to scientific information expressed in these terms,no columns have been provided on the telemetry form. A few words could be fitted in at the bottom of the sheet.

When the form is complete, please return to :

> Project Australis(Telemetry),
> UnionHouse,
> University of Melbourne,
> Parkville,Victoria,
> Australia. 3052

4.9 Station Details

Stations tracking Australis OSCAR A are requested to supply the following information about their station;

(a) name and postal address of operator

(b) call sign or station identification

(c) station latitude and longitude

(d) brief description of v.h.f. equipment such as antenna, pre-amplifier, converter and receiver

(e) brief description of h.f. equipment

(f) brief description of method used to decode the telemetry

Please send these details to the above address, and send emended information whenever a major change is made in your equipment, the date of the change.

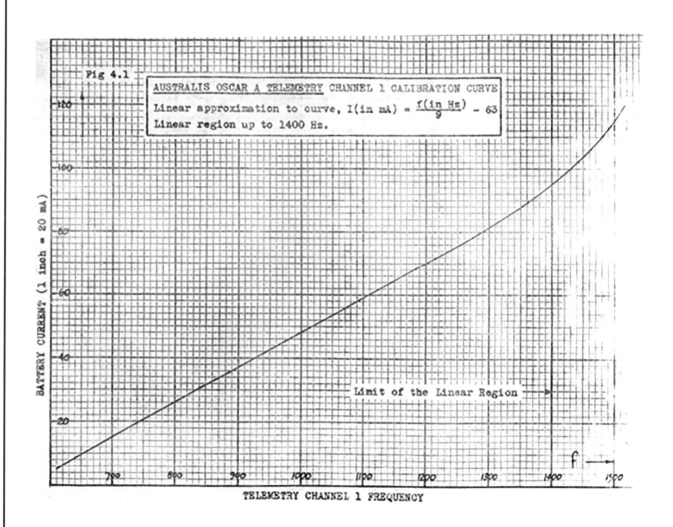

Fig 4.1

AUSTRALIS OSCAR A TELEMETRY CHANNEL 1 CALIBRATION CURVE

Linear approximation to curve, $I(\text{in mA}) = \frac{f(\text{in Hz})}{9} - 63$

Linear region up to 1400 Hz.

TELEMETRY CHANNEL 1 FREQUENCY

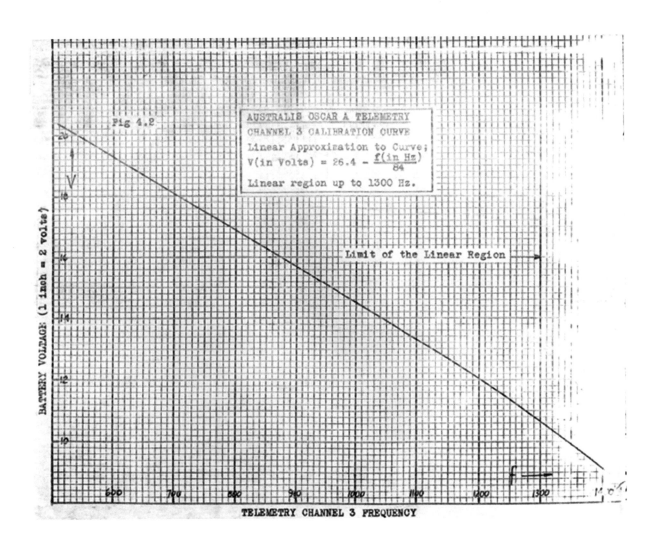

Fig 4.2

AUSTRALIS OSCAR A TELEMETRY
CHANNEL 3 CALIBRATION CURVE
Linear Approximation to Curve;
$$V(\text{in Volts}) = 26.4 - \frac{f(\text{in Hz})}{84}$$
Linear region up to 1300 Hz.

Limit of the Linear Region

BATTERY VOLTAGE (1 inch = 2 volts)

TELEMETRY CHANNEL 3 FREQUENCY

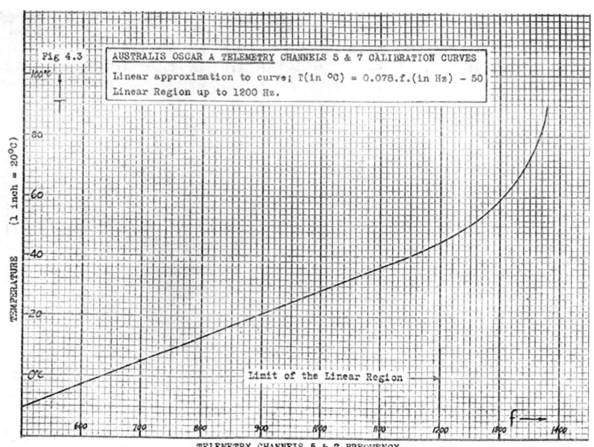

Fig 4.3 <u>AUSTRALIS OSCAR A TELEMETRY CHANNELS 5 & 7 CALIBRATION CURVES</u>

Linear approximation to curve; T(in °C) = 0.078.f.(in Hz) - 50
Linear Region up to 1200 Hz.

TEMPERATURE (1 inch = 20°C)

Limit of the Linear Region ⟶

f ⟶

TELEMETRY CHANNELS 5 & 7 FREQUENCY

Acknowledgments:

Project Australis gratefully acknowledges the kind assistance of the following organisations. Without their help the construction of the satellite would not have been possible.

Acme Engineering, Melbourne - RadioFrequency Connectors

Cannon Electric Ltd., Melbourne - ResistorsandConnectors

Ducon Condensers Pty.Ltd.,Sydney - All Capacitors used in the satellite.

Fairchild Australia Pty Ltd., Melbourne - All Semi-conductors used.

Melbourne University Union - A generous grant for ground equipment

Plessey Components Group, Sydney - One travel grant

The Potter Foundation, Melbourne - Travel Grants for two person to Project OSCAR

Pye Pty. Ltd., Melbourne - All Radio Frequency Crystals

Rola Co. (Aust) Pty. Ltd., Melbourne - MASS magnet

Sample Electronics, Melbourne - Circuit Boards

Turner Industries Ltd., Melbourne - Satellite antennae.

Union Carbide Australia Ltd., Melbourne - Flight and Backup Battery packs

Wireless Institute of Australia - A generous grant for running expenses

Thanks are also extended to the Meteorology Department of Melbourne

University and the Bureau of Meteorology, Melbourne, who have been most helpful during the construction of the satellite.

Australis-Oscar 5 Spacecraft Performance

Note From The Transcriber

This report was transcribed by me from a poor copy of Jan King's QST magazine article of December 1970. I have tried to transcribe the report as closely to the original as I can, including American spelling and two-column layout. The layout differs, of course, for a number of reasons. Tables II, III and IV have not been reproduced. Rendering by other software will no doubt change the layout.

Importantly, it includes references to other articles and papers.

Australis-Oscar 5 Spacecraft Performance

BY JAN S. KING,[1] W3GEY

1   c/o Amsat, P.O. Box 27, Washington, D.C., 20044.

In the rather brief lifetime of the Australis-Oscar 5 experiment a number of useful experimental and operational results have been achieved. The satellite was launched on January 23, 1970. As of this writing, 211 formal reports have been received from 27 countries around the world on both telemetry and propagation results. Many other stations were known to have received the satellite, but did not submit quantitative data.

Based on reports received, here is a summary of the performance of each system on the AO-5 spacecraft.

Thermal Behavior of AO-5

The temperature of AO-5 at ejection from the second stage of the Delta vehicle was 20 degrees C despite its proximity to the second stage engine and a very cold nitrogen gas jet during launch. The temperature, however, began to rise during orbits 1 through 10 and then stabilized internally at 43 degrees C ±3 degrees C where it remained for the duration of the satellite's useful life. This temperature is fairly high, although it is within the design temperature range of 19 degrees to 45 degrees C. The effects of the higher temperature were, unfortunately, all adverse. Battery lifetime was somewhat shortened during the initial phase of discharge, but worse than this, the 144.05 MHz beacon power dropped off faster with decreasing supply voltage due to the decreased efficiency of the power output transistor.

External temperature measurements were higher in the sunlight and cooler during eclipse periods as observed by many responding stations. As the spacecraft entered the dark portion of the orbit the skin temperature dropped from its 55 degrees C average to 42 degrees C ±3 degrees C. The internal temperature, however, remained fairly constant, dropping only two or three degrees during the entire eclipse period. Acknowledgement is due Bill Armstrong, W0PGP, John Fox, W0LER, Nastar, K2SS and others for their data in this area. The spin rate about the X-axis in later orbits became quite slow so that the skin sensor located on the +Y surface showed changes in temperature as parts of the satellite rotated in and out of its own shadow. This data was most useful in determining the roll rate about the stabilized axis of the spacecraft. John Goode, W5CAY, reported this data for many orbits between 100 and 250. Skin temperature data indicated a spin period of 7 to 8 minutes about the X-axis after the initial 100 orbits. An example of this data is shown in Fig. 1 for orbits 168, 205 and 206 along with horizon sensor data.

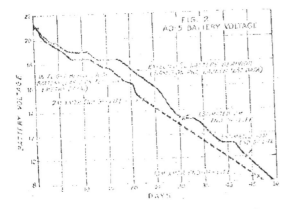

The AO-5 Power System

The spacecraft battery voltage decreased with time faster than predicted by pre-launch testing of individual cells (see Fig. 2). It is now known that the accelerated battery discharge was caused by two factors. First, the higher satellite temperature accelerated the normal chemical reaction in the alkali manganese batteries. Secondly, an additional 18 mA of current was attributed to a failure of the ten-meter modulator that occurred on orbit 3. It was verified that the 18 mA was independent of the ten-meter transmitter itself by commanding the transmitter OFF and observing that the extra current was still present. The ten-meter modulation failure has also been attributed to the higher spacecraft temperature.

The Magnetic Attitude Stabilization System and the Horizon Sensors

One of the best operating systems on board the satellite was not electronic in nature. The Magnetic Attitude Stabilization System (MASS) functioned more efficiently than some of us had anticipated. Early reports indicated that antenna nulls were occurring on the 144.06 MHz signal once every 15 seconds, making telemetry decoding very difficult. By orbit 100, signal fades had reduced to one or two per station pass (approximately 20 minutes in duration). To the station using the spacecraft this is a significant improvement over past satellites in the Oscar series and should prove to be a valuable tool in future amateur spacecraft to achieve the continuous reception of a down-link signal.

The three orthogonal earth or horizon sensors used in the spacecraft were 2N2452 photo transistors operated in diode mode, having a spectral response between 5400 and 10,500 Å. Each sensor's field of view had been stopped to 5 degrees by a small collimation tube. A photometric calibration of those sensors was unfortunately not undertaken due to the shortage of time in the test schedule. While the original design of this part of the telemetry system was to give an ON-OFF indication when looking toward or away from the bright earth, the devices were found to be more sensitive and capable of detecting the decreasing brightness of the earth's atmosphere as the sensors viewed the earth-to-space transition. When viewing the bright earth the telemetry output indication was approximately 1450 Hz and during the transition the telemetry frequency gradually decreased to a dark condition of 600 Hz. Amateurs using a fast discriminator to decode the modulation observed during periods of good signal strength small variations in the frequencies of the telemetry tones as the sensors swept across the earth's disc. These were attributed to cloud formations. Two examples of this data are shown in Fig. 3.

With a discriminator of this type, the Goddard Amateur Radio Club, WA3NAN, decoded telemetry information for all passes received. Figure 4 shows horizon sensor information for various passes. Each frame shows the maximum rate of change of brightness observed on any of the sensors during a given pass. During orbit 4 the maximum observed rate of frequency change was found to be 700 Hz per second, while pass 192 exhibits a maximum rate of change of only 10 Hz per second. This is indicative of the reduced spin rate of the satellite.

During daytime ascending nodes, after the spacecraft had stabilized, a regular sensor pattern was observed. W5CAY demonstrated this data most effectively (see again Fig. 1). The X-axis shows no true periodic nature, but rather a gradual transition followed by small variations about an average "light" condition. The Y and Z sensors show a periodic behavior characteristic of the satellite's roll rate about the stabilized X-axis. The skin temperature shows a cyclic variation as the +Y face rotated in and out of the spacecraft's own shadow. Of particular significance is to observe that the Z sensor always lags behind the Y sensor (approximately two minutes) in detecting the earth. With the X-axis pointing north as the satellite crossed the equator, the spacecraft spin was thus clockwise as observed from the north pole of the earth.

The maxima in the external temperature curve were (within experimental error) out of phase with the +Y sensor. Since the T_{EXT} thermistor was located on the +Y face, then the temperature was a minimum during times when the +Y face was viewing the earth. This is, in fact, the time when the +Y face would have been in shadow.

As the spacecraft traveled north from the equator the +X axis should have begun to dip toward the earth as the strong dipole moment of the satellite (11,800 ???) followed the local geomagnetic field vector which caused it to rotate twice per orbit. (see Fig. 5). W5CAY's data showed that the +X sensor did begin to gradually come on shortly after the signal acquisition time over a period of several minutes. This is precisely what one would have predicted as the +X sensor looked deeper into the earth's atmosphere which reflected more and more scattered light into the sensor.

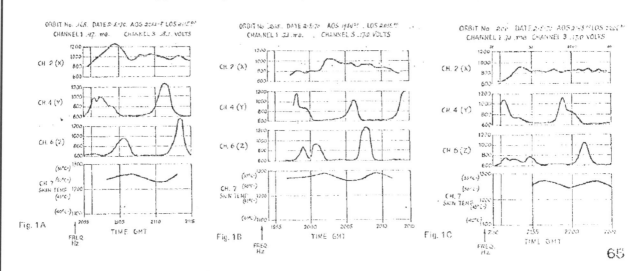

Fig. 1A

Fig. 1B

Fig. 1C

The average roll period observed in this data is 7.5 min. this is thought to be the degree of stabilization that persisted until the termination of the satellite's active life. The effectiveness of this system is best evaluated in terms of the very large reduction in the signal fading rate due to antenna nulls. This, in turn, implies an overall reduction in the loss of spacecraft data. For a satellite in the amateur radio service it is apparent that this method of stabilization is most effective and very easily implemented.

TABLE I

Region	Stations Reporting Useful Data	Stations Reporting Telemetry > 50% Passes	Stations Reporting Telemetry < 50% Passes
I	66	52%	48%
II	114	32%	68%
III	31	45%	55%

The A0-5 Command System

A telecommand link on two meters was utilized to turn ON and OFF the ten-meter beacon transmitter in an effort to conserve the spacecraft's power supply. An a-m tone modulation technique was employed. The ten-meter beacon which consumed 0.6 W of power, was to be commanded ON during the weekends when a maximum number of users was anticipated.

Prior to launch considerable difficulty was encountered with the spacecraft command receiver due to in-band interference from the 144.05 MHz beacon transmitter. It was only possible to eliminate the interference by adding a steep skirted bandpass filter centered at the command frequency. This filter gave 50 dB of rejection at the beacon frequency, but unfortunately had a relatively high insertion loss when placed in front of the receiver. The result was that the command receiver required a signal of -76 dBm (35.4 µV) under ambient (room) conditions to decode a command. This, to be sure, was considered marginal performance. The problem was further complicated

by the detuning of the second i-f stage that occurred during tests under vacuum conditions. The problem could not be traced to a single component in a timely fashion so it was decided to peak the receiver for maximum sensitivity under vacuum conditions. When the receiver was again tested under vacuum conditions the sensitivity was observed to be 10 dB better. Thus, it was expected that the in-flight sensitivity would improve some 10 dB over its ambient condition, giving a final sensitivity figure required to operate the decoder of -86 dBm. The spacecraft was launched with the receiver in this condition.

Figure 6 shows a plot of the spacecraft total current during the entire lifetime of the two-meter beacon, when telemetry data could be obtained. From this data it is clear when commanding occurred and the status of the ten-meter beacon during the lifetime of the satellite. Table III lists the command transmitter schedule indicating the successfully transmitted commands and the effective radiated power used to execute the command. Although early command attempts were unsuccessful, after orbit 72 it became increasingly less difficult to achieve a successful command and it became possible to maintain the weekend-only operation schedule for the ten-meter beacon as originally planned. It is felt that the increased overall sensitivity of the command system was due to a combination of factors:

a) Spacecraft command antenna orientation favorability (particularly over Australia due to the effectiveness of the magnetic attitude stabilization system.)

b) Reduction of the interfering signal level (144.05 MHz) as the battery voltage (and hence the power of the beacon) decreased.

c) Stabilization of the command receiver temperature and pressure which improved the sensitivity of the receiver.

The effectiveness of the command system, particularly despite the receiver problems, is of particular significance to future amateur space experiments. It not only demonstrated, for the first time in an amateur satellite, the effectiveness of ground command as a means of switching various experiments ON and OFF, but of greater significance, it represents an effective means of controlling amateur spacecraft emissions as to prevent interference to other services who may share amateur bands. This should help assure the continuing usage of amateur space experiments without the need for power flux limitations imposed on the spacecraft down-link.

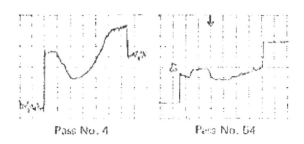

Pass No. 4   Pass No. 54

Fig. 3 – Two examples of variations in the +Y sensor output due to variations in the earth's brightness. Note the sudden increase and decrease in intensity during the frame from pass 54. This is thought to be due to the sensor sweeping across a bright cloud region. Time divisions are 1 sec.

Spacecraft Lifetime

As previously indicated, the failure of the ten-meter modulator is considered responsible for the increased battery current drain of 18 mA. This additional current drain shortened the lifetime of the satellite. The two-meter beacon would be received through approximately orbit 280 on the 23rd day after launch. The ten-meter beacon was turned ON by command on orbit 261 and was left on continuously until it reached end of life around orbit 560 on the 46th day after launch. The difference in beacon lifetimes is due to the variation in cutoff voltage for the transmitters. The two-meter transmitter power output went to zero very rapidly at a supply voltage of 15 V while a significant output could be obtained from the ten-meter transmitter even at voltages as low as ten volts. While the spacecraft lifetime on two meters was shorter than the design lifetime of thirty days, a significant quantity of telemetry data was obtained never the less.

The Nature and Reliability of Amateur Reports

An additional feature of the AO-5 experiment was the opportunity to evaluate the performance of amateurs in reporting scientific-type data. After allowing several months to be certain that all reports had been received, an effort was made to determine what type of information amateurs were most interested in reporting and approximately how much variation in measurement occurred from station to station. It was decided to report on the results by ITU regions since different satellite passes were common to these regions (i.e., Region I (Europe and Africa)) could generally not hear the same passes as Region II (North & South Americas) and so forth). Table 1 lists the number of useful reports received from each region and those which did and did not contain telemetry information. We may infer that stations not reporting telemetry results were primarily interested in other aspects of the experiment or in phenomena such as Doppler measurements. (Only telemetry results are covered in this report since they were the primary indicator of the spacecraft performance. Another report by Raphael Soifer, K2QBW, gives a detailed presentation of the ionospheric propagation results of AO-5. Table 1 indicates that, on a percentage basis, Region

I and Region III participated more actively in the telemetry decoding activities. This is somewhat surprising, since it was anticipated that US amateurs would be suitably equipped to make telemetry measurements.

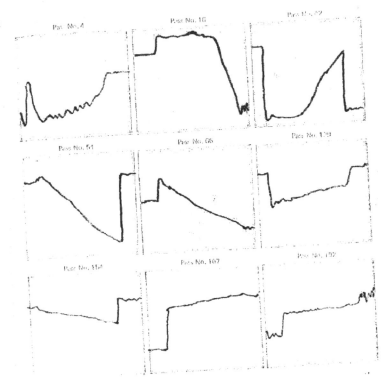

Fig. 4 The maximum rate of change of the horizon sensors during limb transition for various passes of AO-5. The data shows a despin factor of 70 in only 15 days. This is a particularly graphic demonstration of the effectiveness of the stabilization system. Time divisions are 1 sec.

It was of interest to determine the variation in measured values from as many stations as possible during a single pass. Variation in spacecraft parameters for a short period when the satellite passed over a region, was thought to be quite small (except for skin temperature variation) during daylight passes. The variation in data from reporting stations, then, can be primarily considered as individual station measurement error. In each region a particular pass was chosen for which a maximum number of reports was received.

Table II shows data for each station reporting and the range in data as well as the maximum percent of error from the median value. The error observed for the spacecraft battery shows the lowest error due to the relatively "flat" nature of the voltage-to-frequency conversion curve and the fact that most of those reporting rounded off their reported measurement (as called for by the telemetry reporting form). Certain stations (those underlined) were used as control stations for each region since they were known to have better than average decoding equipment.

All regions show comparable data error. The magnitude of the error (less than 10% max.) was approximately the error estimated prior to the launch. This data does not utilize more powerful statistical methods that could be used to more accurately evaluate the data (i.e., a uniform probability density was assumed for all data). The maximum error of 10% does indicate that amateurs throughout the world are capable of making significant data measurements with considerable accuracy.

Summary

With the exception of a failure in the modulator of the ten-meter beacon transmitter, all Australis-Oscar 5 mission objectives were met.

a) The spacecraft was effectively stabilized to two revolutions per orbit (geomagnetic alignment) within the lifetime of the satellite.

b) Reliable amateur spacecraft telecommand was demonstrated.

c) The effectiveness of the seven channel telemetry system was verified. Amateur data generally showed less than ±10% variation from median values.

d) Significant results were obtained on propagation effects over the satellite-to-earth link in the ten-meter band.

e) Partial success was obtained in achieving the design lifetime of several weeks for both spacecraft transmitters using only chemical batteries.

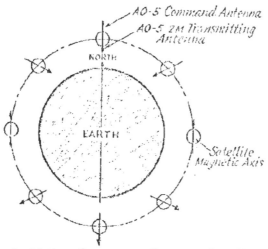

Fig. 6 – *Motion of a magnetically oriented satellite in a polar orbit.*

While the response to A0-5 was gratifying (many stations reported it to be the most interesting amateur space activity to date) it does not compare with the level of excitement that was generated by the repeater satellites such as Oscar III. Amsat is presently planning a next generation of Oscars. These satellites will carry two repeaters and an RTTY telemetry system capable of measuring as many as 60 parameters. The design lifetime of these satellites will be one year using a solar cell power source.

Whether you are interested in RTTY, fm, a-m, ssb, DX, traffic handling, or even contesting there are activities and special experiments being planned for you with Oscar 6. If you are interested in finding out how you can contribute to this new and exciting chapter in amateur radio, write to Amsat, P.O. Box 27, Washington, D.C., 20044.

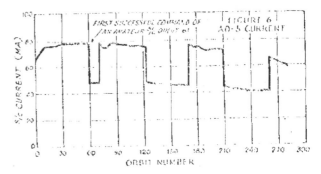

Bibliography

Data taken from a series of reports on Australis-Oscar 5 submitted to Amsat by John Goode, W5CAY.

1. Data taken from Australis-Oscar 5 (A Summary Report) submitted to Amsat by John Fox, W0LER.
2. Data taken from Fairchild Semiconductor Specification information on 2N986/2N2452 NPN Planar Phototransistor, 5/62.
3. Information from preliminary data reduced at the Goddard Space Flight Center, NASA, by the Goddard Amateur Radio Club, 4/70.
4. Fischell, Robert E., "Magnetic and Gravity Attitude Stabilization of Earth Satellites," report CM 996, John Hopkins Univ. Applied Physics Labs, May 1961, p.38.
5. Op. Cit., John Fox, W0LER.
6. Soifer, Rafael, "Ionospheric Propagation from Australis-Oscar 5" (A Survey Report to the Radio Amateur Satellite Corporation). QSL October, 1970, pg. 54.
7. Op. Cit., Soifer.

Rocket, TIROS M and Orbit

Launch Vehicle

The launch rocket was a two-stage, long tank, thrust augmented Thor/Delta rocket[1]. In addition, there were six solid propellant, strap-on Thiokol Castor II motors used for the first time in this configuration with a Thor/Delta. Interestingly, three of the Castors were ignited thirty seconds after lift-off. The lift-off thrust was about 270 tonnes (nearly 600,000 pounds) and was capable of orbiting a satellite of twice the mass of the satellite it launched, TIROS M. Updated and improved versions are today manufactured and flown by the United Launch Alliance.

TIROS M Satellite

The TIROS M[2], also known as the Improved TIROS Operational Satellite (ITOS), was a cylinder 1.2 metres high, a span of 4.3 metres over its solar panels, and it weighed 330kg at launch.

The following instruments were carried: an Advanced Vidicon Camera System, an Automatic Picture Transmission (APT) system, a Flat Plate Radiometer, a Solar Proton Monitor, a Vertical Temperature Profile Radiometer and a Very High Resolution Radiometer. It provided data for nearly 18 months, ending its life on 18 June 1971.

The Orbit

We considered that the orbits of TIROS and AO5 were identical if for no other reason than that for several weeks we had no other means of knowing the orbit of AO5. In any event, it was a perfectly satisfactory assumption for the purpose of providing antenna pointing information for radio amateurs.

Apogee (highest point in the orbit):	1,479 km
Perigee (lowest point in the orbit):	1,434 km
Inclination:	101.8°
Period:	115 minutes exactly

---

1   See https://en.wikipedia.org/wiki/Thor-Delta; https://en.wikipedia.org/Thor_(rocket_family); and www.historicspacecraft.com/Rockets_Delta.

2   See http://www.astronautix.com/i/itos.html; and https://www.wmo-sat.info/oscar/satellites/view/203.

APPENDIX D

Balloon Yarns

As promised, this appendix records some of the funniest and silliest ballooning incidents, giving awards to each. Please don't think that disaster was the norm as most flights were successful.

However, to improve the storytelling, there might be a slight amount of exaggeration herein!

The Aviation Award for the World's Largest Party Balloon

The launch was quite normal and the balloon rose as majestically as ever while we waited for it to reach float altitude and we could turn on our instrument. It was not too long before word came from the flight crew that it seemed to be climbing more slowly than expected, then not at all. It was a leaker and floating at, oh dear, an altitude frequented by commercial jets. Not a problem, we can cut it down. I could almost hear the bang of the rope cutters.

Not really; there was an overriding pressure switch that prevented cut down until it reached a much higher altitude. Commercial jets would simply have to look out for the balloon as there was no way to destroy it. Never mind, we were told, it will eventually come down in a few days – or weeks. It was the latter.

After carefully following assigned commercial aviation routes across the country and slowly descending, the balloon eventually touched the ground. Or at least our payload did and the wind dragged it across the ground until a large piece of our instrument broke off. Up, up and away again into the commercial air routes once more. Up and down a few times until our instrument was totally destroyed and the balloon's buoyancy was gone.

Override pressure switches were no longer used after that flight.

The Tex Award for Optimism

I arrived at the National Scientific Balloon Station, Palestine, Texas, one day to be told that there was to be a test next morning of a balloon inflation with hydrogen gas. Helium is normally used to inflate balloons but supplies are limited to a much greater extent than hydrogen and, besides, hydrogen is cheaper than helium. Conveniently for NSBF, however, helium issues from the ground a hundred or so kilometres away near Amarillo, Texas.

On the negative side for hydrogen, the public associates the Hindenburg disaster in May 1937 with explosions, fire and disasters due to hydrogen. The fact that most, but not all, who died on the Hindenburg did so as a result of jumping from the airship seems to have escaped notice, and many were burned. The days of hydrogen airships were over and helium became the preferred gas for buoyancy.

An old balloon to be used for the test was rolled out and inflation began with the crew dressed in white fire suits and helmets. Once the test had shown that it was possible to inflate the balloon without it spontaneously catching fire, it was time to destroy it. Two cut-down explosive cutters were installed at the top of the balloon, but no one had thought to attach the cut-down cord to something firmly fixed to the ground – you can see what's coming, can't you?

Well, the first explosive cutter was set off and nothing happened – gas escapes ever so slowly from a miserable little hole in the balloon. No fire, as was hoped. Bang, same result with the second.

Never mind, Tex is with us – he will solve the problem. So Tex takes out his two-pearl handled Colt 45s from his holster (literally) and starts firing away. Unsurprisingly, these had even less effect on the balloon even after emptying his two pistols of bullets.

What to do now? Well, as originally planned, the hydrogen needs to be set alight to destroy the balloon. Easy – a lighted taper will do the trick. Trouble is, the balloon is many metres above and we want to avoid an explosion; after all, the Hindenburg exploded, didn't it? (No.) So, a towel soaked in petrol tied to a broom will do the trick, won't it? No, we'll need more than one broom, so two were tied together. Still not enough, so back to the headquarters building for another broom. And another. Does this remind you of the Marx brothers or Monty Python?

Eventually, the burning towel atop of a row of tottering brooms reached the balloon and it was set on fire. The remains? Great gobs of balloon plastic on the tarmac and a number of red Texan faces. Yes, it is possible to fly hydrogen-filled balloons (all weather balloons are, for example) and it can be done safely.

And don't forget the cut-down cord – it's there for a reason.

The RAAF Award for Low Flying Aircraft

Another perfect launch and flight except for a minor detail. The cut-down command didn't. No worries, that gives us more time to collect data.

Over a few days, the helium gas leaks from a balloon, so it slowly descends. Naturally, this one chose Sydney for its descent into the aviation routes before heading, fortuitously, towards RAAF Base Williamtown where our front-line fighters[1] were based. (Yes, it's an odd place to base front line fighters, but perhaps the RAAF bigwigs were preparing for the day that a massive balloon attacked their base.)

What is more natural for fighter aircraft than to shoot down an enormous balloon?

Having read the previous award, you can guess, dear reader, the success of the enterprise.

---

1   In those days, they were French Mirage fighters.

I gather that one pilot brought his aircraft dangerously close to the balloon and narrowly missed winning an ignominious award for being the only Mirage pilot ever to be brought down by a balloon.

The Broken Hill Award for the World's Largest Dry Cleaning Cover

Perfect launch conditions at Broken Hill and a perfect launch. Well, a near-perfect launch. Off went the launch truck with our instrument. And off, and off, and off. Release the damned thing, we kept saying to ourselves. Off went the launch truck until it reached the airport's boundary fence. It was not promising territory beyond the fence: a road and steep cliff to a railway line.

Oh well, just cut it down. Bang! And the balloon launched itself, ripping apart as the cut-down cord did its job and the balloon was destroyed. As expected, the wind carried the balloon over the road and over the railway line. That is, acres of balloon plastic covered the road and line.

Try stuffing the remains of a balloon back into its box! Quickly before the morning freight train arrives.

A Conspiracy Theory

Surely no book like this is incomplete without a conspiracy theory? Did the Commonwealth decide to go ahead with the WRESAT satellite in order to beat the students' satellite into orbit? Was there Australian pressure on the US Air Force to falter in its cooperation with Project OSCAR? The answer to both questions is almost certainly not, but …

Before examining any conspiracy theory, let us establish that Australis was indeed the first satellite built in Australia. To quote Des Barnsley, the project manager for WRESAT[1], 'Towards the end of 1966 it was clear that at least one of the 10 SPARTA vehicles would be surplus to project needs and the possibility of using it to launch an Australian designed satellite was unofficially raised. This idea quickly took off and agreement between the Department of Supply and the US DoD was speedily obtained.' Later in the referenced web page Barnsley states that 'Design work on WRESAT began at W. R. E. and the University of Adelaide early in 1967 …'

Australis was delivered to the US in June 1967 while WRESAT was launched on 29 November 1967, soon after its tests had been completed. Australis was completed some five months before WRESAT and therefore earned the title of Australia's first satellite but not the first to orbit.

I have a number of newspaper clippings stating that we expected a launch of Australis within 12 months of being delivered to the US. It is ironic that WRESAT was launched before Australis.

I have a letter from the Minister for Air dated 4 March 1966 acknowledging my advice to him of our 'proposed satellite'. Thus the Australian government was aware of Australis nearly a year before WRESAT was approved and design work on it began. It is unlikely that the Minister for Air would not have mentioned Australis to the Minister for Supply who was responsible for WRE. However, I can find no evidence that WRESAT was approved in order to trump Australis nor have I been able to find any mention of WRESAT in the cabinet papers of the day.

Nonetheless, suggestive information, but not definitive proof, is the stock-in-trade of conspiracy theorists.

---

1   'WRESAT Background' at <https://www.honeysucklecreek.net/supply/WRESAT/background. html>. Accessed July 2017.

Image Credits

All images have been taken by the me and other members of the Australis OSCAR team with the exception of the following:

Plate 1.1: Credit: NASA.

Plate 2.1: Image Credit: Bauer Media Pty Limited.

Plate 2.3: Permission of Arc @ UNSW Ltd.

Plate 7.4: Credit: Fairfax Syndication.

Plate 10.5: Credit: NASA.

Plate 10.9: Reprinted with permission from the University of Melbourne.

Plate 10.10: Credit: NASA.